Quick Reference to Useful Design Aids

SIMPLIFIED DESIGN OF BUILDING FOUNDATIONS

Simplified Design of Building Foundations

JAMES AMBROSE
Professor of Architecture
University of Southern California

A WILEY-INTERSCIENCE PUBLICATION
JOHN WILEY & SONS
New York • Chichester • Brisbane • Toronto

This publication is designed to provide accurate and authoritative information in regard to the subject matter covered. It is sold with the understanding that the publisher is not engaged in rendering legal, accounting, or other professional service. If legal advice or other expert assistance is required, the services of a competent professional person should be sought. *From a Declaration of Principles jointly adopted by a Committee of the American Bar Association and a Committee of Publishers.*

Library of Congress Cataloging in Publication Data:

Ambrose, James E
Simplified design of building foundations.

"A Wiley-Interscience publication."
Includes index.
1. Foundations–Design and construction.
I. Title
TH2101.A58 690'.11 80-39880
ISBN 0-471-06267-7

Printed in the United States of America

10 9 8 7 6

Preface

This book is intended as an introductory text for courses in the design of building foundations and as a guide for the practical design of simple foundation elements and systems. As the title implies, the work is limited in scope and intensity. A minimal background preparation by the reader is assumed, consisting of an introductory study of applied mechanics (statics and strength of materials) and the design of simple structural elements of masonry and reinforced concrete. Mathematical work is limited to algebra, plane geometry, and elementary plane trigonometry.

Although the book is intended primarily for use by students in architecture and building technology programs, it should also be useful as an introduction to the general topic of building foundations for students in civil engineering and structural engineering programs. The comprehensive presentation of the subject given here will provide an excellent overview and orientation for the reader who intends to pursue studies of a more intensive nature.

I am grateful to John Wiley & Sons, the American Concrete Institute, Building News, and the International Conference of Building Officials for permission to reproduce materials from their publications. I am indebted to my colleague Dimitry Vergun for his careful reviews of the manuscript and his many helpful suggestions. Finally, I am grateful to my family for their patience and support.

Los Angeles, California
February 1981

JAMES AMBROSE

Contents

building designer (architect), the consulting structural engineer, the soils engineer or testing service, and the builder.

Architect. The architect usually has overall responsibility for all of the design work for the building and sometimes for inspection of the construction. The architect is usually the one who retains the services of the consulting engineers and coordinates their work with the general development of the design and the construction. The architect must take care to seek expert advice at the proper time so that the design work does not proceed too far before the advice is given.

Some approximate determination of the building size and form must be done by the architect before the structural engineer can determine the foundation loading conditions. The position of the building on the site and the general site design must also be known before decisions can be made about the foundation. On many projects the site design work may also involve the use of specialists such as landscape architects and civil engineers.

It is often necessary to allow for some sequential interchange between the work of the various persons involved in the design. The building design and site design cannot be carried too far without some input from the site investigation and advice from the soils consultant. However, the final detailed subsurface investigation should preferably not be done until information is available about the building location, the nature and magnitude of the foundation loads, the final site grading, and so on. Figure 1.1 illustrates an idealized sequencing of the various activities proceeding from the first design work to the beginning of the construction.

Although the sequence shown in Figure 1.1 may be ideal from a design point of view, the time required to design and construct large projects often favors the use of a fast-tracking operation, in which the construction is begun before the design work is complete. Figure 1.2 illustrates such a situation, in which the foundation construction is begun before the final details of the building construction are fully determined. Such a situation often requires a slightly conservative design for the foundation, since changes in the final building design could result in problems for the already-built foundation.

Structural Engineer. Bearing in mind the preceding discussion of the responsibilities of the architect and the timing of design and construc-

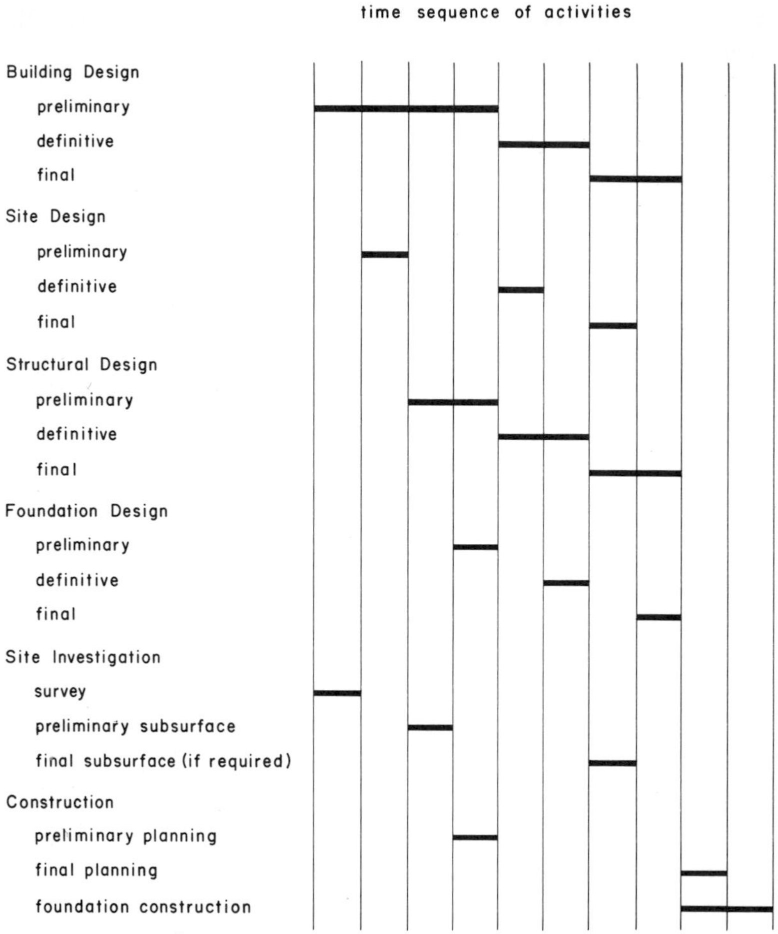

FIGURE 1.1. Typical sequencing of building design and construction activities.

tion activities, the following are some of the usual specific functions of the structural designer, whether he or she is an outside consultant or an employee of the architect.

1. Determining the loads on the foundation.
2. Developing the specifications for the subsurface investigations.
3. Interpreting the site investigation reports and giving advice to the architect on the general feasibility of the building and site designs.

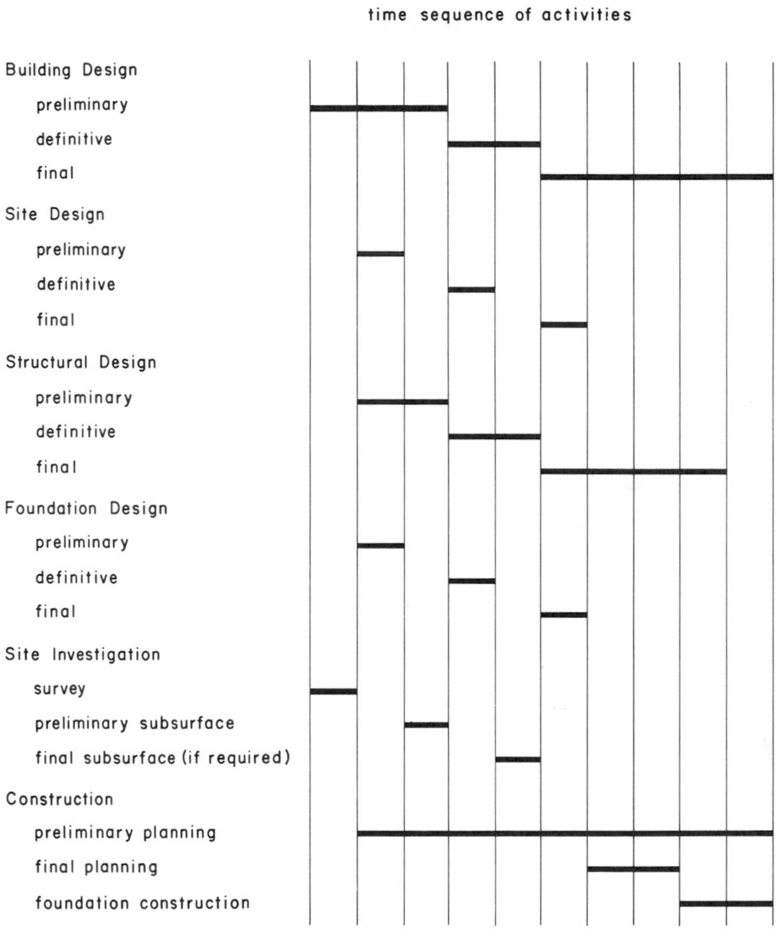

FIGURE 1.2. Sequencing of activities for a fast tracking of building design and construction.

4. Deciding on the type of foundation required and performing of the detailed design of the foundation elements.
5. Conducting inspections required during excavation and construction.
6. Reviewing any tests made during construction and recommending any alterations of the foundation design required because of errors or conditions encountered at the site.

Soils Engineer. In most cases subsurface investigation is done by an organization that performs the site investigation, any required testing, and submits a report with recommendations for the site and foundation designs. The report is usually prepared by an engineer who specializes in foundation problems. Specific functions that may be performed by this service are the following:

1. Performing the investigation and tests.
2. Recommending the type of foundation, allowable soil pressures, use of site materials, and any special provisions required for dewatering, expansive soils, soil stabilization, and so on.
3. Recommending and performing any necessary additional investigations.
4. Performing any special inspections or tests during construction.

Builder. Depending on the size and nature of the building project and the conditions at the site, considerable engineering design may be involved in getting the building built. The division of responsibility for this design between the architect, the structural engineer, the soils engineer, and the builder can become complex. Some situations that may have to be dealt with are the following:

1. Developing the construction timetable, and interfacing with the design timetable if the project is fast tracked.
2. Providing for disposal of excavated materials.
3. Dewatering the excavation during construction of the foundation.
4. Constructing temporary bracing for deep excavations to prevent settlement or undermining of adjacent property.

There are many possible types of contractual arrangements between the various parties involved in the design, construction, and ownership of a building. There are also many persons and groups involved in addition to those previously discussed, such as the building owner, code-enforcing agencies, financing institutions, insurers, unions, and so on. The process of design decision making often involves a large number of influences. It has been our purpose here to explain the major activities involved and the usual functions of the major participants.

1.3 Design of Concrete Foundation Elements

The majority of structural elements used for foundations are made of concrete. A major part of foundation design, therefore, consists of designing various types of structural elements of concrete. It is not the purpose of this book to serve as a text on design of concrete structures in general, although considerable use is made of structural calculations in the illustrative design examples.

In order to keep the work in this book within the scope of simplified engineering, the analysis and design of concrete elements is done with the working stress method rather than with the strength design methods currently favored in engineering practice. It is felt that persons with limited backgrounds in engineering will find this method to be simpler and easier to follow. A brief presentation of the formulas and procedures of the working stress method as applicable to the work shown in this book is given in the appendix. Also included in the appendix are selected reprints from the 1963 edition of the ACI Code (Ref. 11), which was the last edition of this code that contained extensive material of use for the working stress method.

While the author does not recommend use of the working stress method for design of reinforced concrete structures in general, it is felt that the method is reasonably acceptable for the types of elements illustrated in most of the examples in this book. At the present time most agencies that grant building permits will accept this design method for elements limited to the conditions shown in the examples. In specific instances, however, it is wise to verify this before submitting designs for permit consideration.

Justification for use of the working stress method is based on some assumptions regarding the typical conditions that exist in the construction of ordinary foundation elements. The principal assumptions are the following:

1. Concrete strengths are usually low; commonly used are f_c' values of 2000 and 3000. For small building projects 2000 is commonly used because most codes permit the omission of site inspection and testing of the concrete when the f_c' used in design is limited to 2000 psi. For larger projects or heavier loads a higher strength may be needed, although if a strength in excess of 3000 psi is required, it is likely that the structural design in general is beyond the reasonable scope of simplified engineering procedures.
2. Steel reinforcing used in ordinary foundations is usually of a low grade with design stresses correspondingly low.
3. The amount of reinforcing used in ordinary foundations is usually quite low. Shear reinforcement is seldom used and flexural

reinforcing is usually well below the limits of capability of the concrete sections. This is largely a matter of economics; it is generally more economical to use more concrete and less reinforcing.

If these conditions exist, the difference in design results obtained by the use of the working stress and strength methods will usually be minor. However, the use of the latest codes and the current strength design methods is recommended when the design involves any of the following conditions:

1. Use of concrete with f_c' in excess of 3000 psi.
2. Use of steel reinforcing with f_y in excess of 60,000 psi.
3. Use of large percentages of tensile reinforcing, approaching or exceeding the balanced section capacity of the concrete.
4. Use of flexural elements with compressive reinforcing.
5. Designs requiring major amounts of shear reinforcing.

It is necessary, of course, that any design conform to the general requirements of the latest codes. These should be used to determine items such as minimum percentages of reinforcement, cover for reinforcing, minimum sizes of elements, and so on.

Readers who are adequately prepared to utilize strength design methods are encouraged to do so in any actual design work. Those who desire to gain such preparation may find a useful introduction to the method in *Simplified Design of Reinforced Concrete* by Harry Parker (Ref. 3).

1.4 Use of English and Metric Units

We have chosen to use English units in this book primarily because they are still being used at the time of this writing in all of the principal references that have been utilized for the work in the book. This is, of course, a time of transition and most engineering fields are in the process of changing to the new metric units. In due time the building codes, the building industry in general, the standard handbooks, and the basic texts in structural engineering will undoubtedly all be in metric units. For some time to come, however, persons working in this field will have to endure the clumsy work of translating continuously from one system to the other.

1.5 How to Use This Book

Within the limits of its size this book is intended to provide a treatment of both basic principles and practical aspects of the design of foundations for buildings. It is anticipated that the book will be used for self-study or for classes in elementary foundation engineering as well as for a reference guide for beginning structural designers. The book should be equally useful as a complete text for those whose interests are limited to relatively simple building design problems and as an introductory text for those who intend to pursue the study of foundation engineering for buildings to a more professional level.

The scope of discussion is quite broad and the treatment reasonably thorough with regard to the general problems and basic principles of soil behavior and foundation engineering. This is done so that readers can become reasonably familiar with the range of problems encountered in building design and construction. The author favors this approach so readers can orient themselves with regard to their individual capabilities in terms of design skills and knowledge.

Illustrated calculations for actual design solutions are limited to relatively simple situations. These include the design of wall and column footings, combined column footings, basement walls, short cantilever retaining walls, and short piers of masonry and concrete. Soil conditions assumed in the design examples are also limited to relatively ordinary situations. For more rapid access a list of the illustrated design examples is given inside the back cover of the book. Also given there is a list of tables and figures with useful design information.

Readers wishing to pursue the study of more complex foundations and more unusual soil conditions may use the references listed at the back of the book, following Chapter 6. Many of these references are cited throughout the book.

As a test of comprehension, as well as for general skill development, readers are urged to make use of the questions and exercise problems that are provided at the end of some of the chapters. After attempting to solve the exercise problems, the readers may refer to the selected answers, which are given in the back of the book, following the appendix. To keep calculations within the scope of persons with limited backgrounds, design examples utilize the simpler procedures of the working stress method for design of reinforced concrete. A summary of the formulas and procedures for this method, together with reprints of the 1963 edition of the *ACI Code*, are given in the appendix. Readers who are prepared to use strength design methods may use them for the exercise problems, but will obtain answers slightly different from those given in the back of the book.

2

Soil Properties and Foundation Behavior

Information about the materials that constitute the earth's surface is forthcoming from a number of sources. Persons and agencies involved in fields such as agriculture, landscaping, highway and airport paving, waterway and dam construction, and the basic earth sciences such as geology, mineralogy and hydrology have generated research and experience that is useful to the field of foundation engineering. This chapter consists of a brief summary of the issues and data regarding soil materials and behavior that directly concern the designer of building foundations. It will provide the reader with a general understanding of the problems of soil identification and the means for establishing criteria for foundation design.

2.1 Soil Considerations Related to Foundation Design

The principal properties and behavior characteristics of soils that are of direct concern in foundation design are the following:

Strength. For bearing-type foundations the main concern is resistance to vertical compression. Resistance to horizontal pressure and to friction are of concern when foundations must resist the force of wind, earthquakes, or retained earth.

Strain resistance. Deformation of soil under stress is of concern in designing for limitations of the movements of foundations, such as the vertical settlement of bearing foundations.

Stability. Frost action, fluctuations in water content, seismic shock, organic decomposition, and disturbance during construction are some of the things that may produce changes in physical properties of soils. The degree of sensitivity of the soil to these actions is called its relative stability.

Properties Affecting Construction Activity. A number of possible factors may affect construction activity, including the following:

The relative ease of excavation.

Ease of and possible effects of site dewatering during construction.

Feasibility of using excavated materials as fill material.

Ability of the soil to stand on a vertical side of an excavation.

Effects of construction activity—notably the movement of workers and equipment—on unstable soils.

Miscellaneous Conditions. In specific situations various factors may affect the foundation design or the problems to be dealt with during construction. Some examples are the following:

Location of the water table, affecting soil strength or stability, need for waterproofing basements, requirement for dewatering during construction, and so on.

Nonuniform soil conditions on the site, such as soil strata that are not horizontal, strips or pockets of poor soil, and so on.

Local frost conditions, affecting the depth required for bearing foundations and possible heave and settlement of exterior pavements.

Deep excavation or dewatering operations, possibly affecting the stability of adjacent properties, buildings, streets, and so on.

All of these concerns must be anticipated and dealt with in designing buildings and in planning and estimating construction costs. Persons charged with responsibility for design and planning foundation construction must have some understanding of the characteristics of ordinary soils so that they can translate information about site conditions into usable data. The discussions that follow deal with the basic nature of soils of various types, the behavior and design considerations of various foundation elements and systems, and the means for obtaining and using information about specific site conditions.

2.2 Soil Properties and Identification

A general distinction can be made between two basic materials: *soil* and *rock*. At the extreme the distinction is clear, loose sand versus solid granite, for example. A precise distinction is somewhat more difficult, however, since some highly compressed soils may be quite hard, while some types of rock are quite soft or contain many fractures, making them relatively susceptible to disintegration. For practical use in engineering, soil is generally defined as material consisting of discrete particles that are relatively easy to separate, while rock is any material that requires considerable brute force for excavation.

A typical soil mass is visualized as consisting of three parts, as shown in Figure 2.1. The total soil volume is taken up partly by the solid particles and partly by the open spaces between the particles, called the void. The void is typically filled partly by liquid (usually water) and partly by gas (usually air). There are several soil properties that can be expressed in terms of this composition, such as the following:

Soil weight (γ). Most of the materials that constitute the solid particles in ordinary soils have a unit density that falls within a narrow

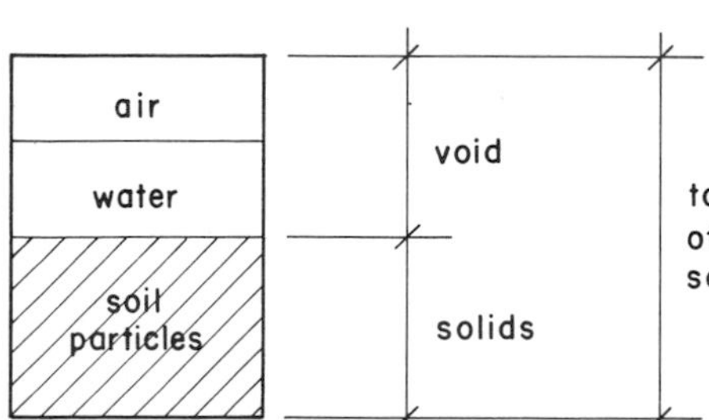

FIGURE 2.1. Three-part composition of a soil mass.

range; expressed as specific gravity, the range is from 2.60 to 2.75. Sands typically average about 2.65; clays about 2.70. Notable exceptions are soils containing large amounts of organic materials. Specific gravity refers to a comparison of the density to that of water, usually considered to weigh 62.4 lb/ft^3. Thus, for a dry soil sample the soil weight may be determined as follows:

soil unit weight = (% of solids)(specific gravity)(unit weight of water)

Thus for a sandy soil with a void of 30%, the weight may be approximated as follows:

$$\text{soil weight} = \gamma = \left(\tfrac{70}{100}\right)(2.65)(62.4) = 116 \text{ lb/ft}^3$$

Void ratio (e). Instead of expressing the void as a percentage, as was done in the preceding example, the term generally used is the void ratio, e, which is defined as follows:

$$e = \frac{\text{volume of the void}}{\text{volume of the solid}}$$

In practice the void ratio is often determined by using the relationship between soil weight, specific gravity of the solids, and percentage of the void, as follows:

if

$$\gamma = \frac{\text{measured dry weight of the sample}}{\text{measured volume of the sample}} = 116 \text{ lb/ft}^3$$

then, assuming a specific gravity (G_s) of 2.65,

$$\gamma = \frac{(\%\text{ of solids})}{100}(2.65)(62.4) = 116 \text{ lb/ft}^3$$

$$\%\text{ of solids} = \frac{116}{(2.65)(62.4)}(100) = 70\%$$

$$\%\text{ of void} = 100 - 70 = 30\%$$

and, with the volume expressed as a percentage,

$$e = \frac{\text{volume of the void}}{\text{volume of the solid}} = \frac{30}{70} = 0.43$$

Porosity (n). The actual percentage of the void is expressed as the porosity of the soil, which in coarse-grained soils (sands and gravels) is generally an indication of the rate at which water flows through or drains from the soil. The actual water flow is determined by standard tests, however, and is described as the relative *permeability* of the soil. Porosity, when used, is simply determined as

$$n \text{ (in percent)} = \frac{\text{volume of the void}}{\text{total soil volume}} (100)$$

Thus for the preceding example, $n = 30\%$.

Water content (w). The amount of water in a soil sample can be expressed in two ways: by the water content (w) and by the saturation (S). They are defined as follows:

$$w \text{ (in percent)} = \frac{\text{weight of water in the sample}}{\text{weight of solids in the sample}} (100)$$

The weight of the water is simply determined by weighing the wet sample and then drying it to find the dry weight.

The saturation is expressed in a ratio, similar to the void ratio, as follows:

$$S = \frac{\text{volume of water}}{\text{volume of void}}$$

Full saturation ($S = 1.0$) thus occurs when the void is totally filled with water. Oversaturation ($S > 1.0$) is possible in some soils when the water literally floats some of the solid particles, increasing the void above that in the partly saturated soil mass.

In the preceding example, if the soil weight of the sample as taken at the site was found to be 125 lb/ft^3, the water content and saturation would be as follows:

$$\begin{aligned} \text{weight of water} &= \text{(original sample weight)} - \text{(dry weight)} \\ &= 125 - 116 = 9 \text{ lb/ft}^3 \end{aligned}$$

Then

$$w = \frac{9}{116} (100) = 7.76\%$$

The volume of water may be found as

$$V_w = \frac{\text{weight of water in sample}}{\text{unit weight of water}} = \frac{9}{62.4} = 0.144 \text{ or } 14.4\%$$

Then

$$S = \frac{14.4}{30} = 0.48$$

The size of the discrete particles that constitute the solids in a soil is significant with regard to the identification of the soil and the evaluation of many of its physical characteristics. Most soils have a range of particles of various sizes, so the full identification typically consists of determining the percentage of particles of particular size categories.

The two common means for measuring grain size are by sieve and sedimentation. The sieve method consists of passing the pulverized dry soil sample through a series of sieves with increasingly smaller openings. The percentage of the total original sample retained on each sieve is recorded. The finest sieve is a No. 200, with openings of approximately 0.003 in. A broad distinction is made between the total amount of solid particles that pass the No. 200 sieve and those retained on all the sieves. Those passing are called the *fines* and the total retained is called the *coarse fraction*.

The fine-grained soil particles are subjected to a sedimentation test. This consists of placing the dry soil in a sealed container with water, shaking the container, and measuring the rate of settlement of the particles. The coarser particles will settle in a few minutes; the finest will take several days.

Figure 2.2 shows a graph that is commonly used to record the grain size characteristics for soils. A log scale is used for the grain size, since the range is quite large. The common soil names, based on grain size, are given at the top of the graph. These are approximations, since some overlap occurs at the boundaries, particularly for the fines. The distinction between sand and gravel is specifically established by the No. 4 sieve, although the actual materials that constitute the coarse fraction are sometimes the same across the grain size range. The curves shown on the graph are representative of some particularly characteristic soils, described as follows:

A *well-graded* soil consists of some significant percentages of a wide range of soil particles.

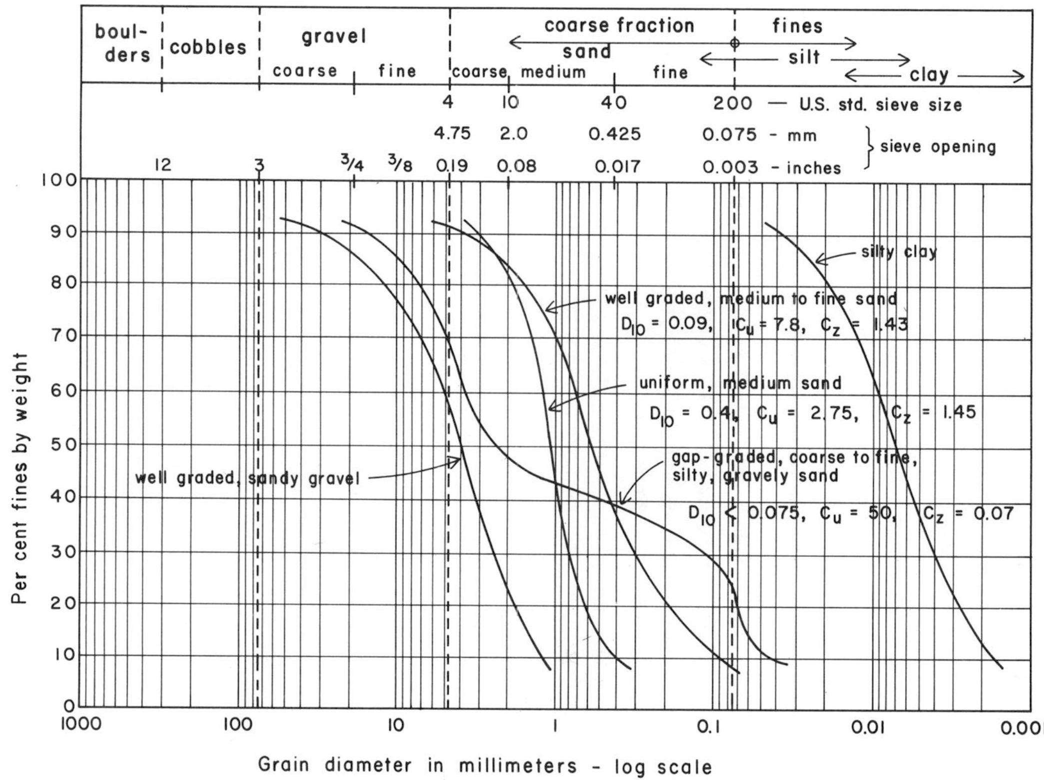

FIGURE 2.2. Grain size range for typical soils.

A so-called *uniform* soil has a major portion of the particles grouped in a small size range.

A *gap-graded* soil has a wide range of sizes, but with some concentrations of single sizes and small percentages over some ranges.

These size-range characteristics are specifically established by using some actual numeric values from the size-range graph. The three size values used are points at which the curve crosses the percent lines for 10, 30, and 60%. The values are interpreted as follows:

Major size range. This is established by the value of the grain size in mm at the 10% line, called D_{10}. This expresses the fact that 90% of the solids are above a certain grain size. The D_{10} value is specifically defined as the *effective grain size*.

Degree of size gradation. The distinction between uniform and well-graded has to do with the slope of the major portion of the size-range curve. This is established by comparing the size value at the 10% line, D_{10}, with the size value at the 60% line, D_{60}. The relationship is expressed by the *uniformity coefficient* (C_u), which is defined as

$$C_u = \frac{D_{60}}{D_{10}}$$

The higher this number, the greater the degree of size gradation.

Continuity of gradation. The value of C_u does not express the character of the graph between D_{60} and D_{10}; that is, it does not establish whether the soil is gap-graded or well-graded, only that it is graded. To establish this, another property is defined, called the *coefficient of curvature* (C_z), that uses all three size values, D_{60}, D_{30}, and D_{10}, as follows:

$$C_z = \frac{(D_{30})^2}{(D_{10})(D_{60})}$$

These coefficients are used only for classification of the coarse-grained soils: sand and gravel. For a well-graded gravel, C_u should be greater than four, and C_z between one and three. For a well-graded sand, C_u should be greater than six, and C_z between one and three.

The shape of soil particles is also significant for some soil properties. The three major classes of shape are bulky, flaky, and needlelike, the latter being quite rare. Sand and gravel are typically bulky; further distinction is made with regard to the degree of roundedness of the particle form. Bulky-grained soils are usually quite strong in resisting static loads, especially when the grain shape is quite angular, as opposed to well rounded. Unless a bulky-grained soil is well-graded or contains some significant amount of fine-grained material, however, it tends to be subject to displacement and consolidation due to vibration or shock.

Flaky-grained soils tend to be easily deformable and highly compressible, similar to the action of randomly thrown loose sheets of paper or dry leaves in a container. A small percentage of flaky-grained particles can impart the character of a flaky soil to an entire soil mass.

Water has various effects on soils, depending on the proportion of water and on the particle shape, size and chemical properties. A small amount of water tends to make sand particles stick together somewhat. As a result, the sand behaves differently than usual, no longer acting as a loose, flowing mass. When saturated, however, most sands behave like viscous fluids, moving easily under stress due to gravity or other sources. The effect of the variation of water content is generally more dramatic in fine-grained soils. These will change from rocklike solids when totally dry to virtual fluids when supersaturated.

Table 2.1 describes for fine-grained soils the Atterberg limits, which are the water content limits, or boundaries, between four stages of structural character of the soil. An important property of such soils is the *plasticity index*, I_p, which is the numeric difference between the liquid limit and plastic limit. A major physical distinction between clays and silts is the range of the plastic state, referred to as the relative plasticity of the soil. Clays have a considerable plastic range and silts generally have practically none, going almost directly from the semisolid to the liquid state. The plasticity chart, shown in Figure 2.3, is used to classify clays and silts on the basis of two properties, liquid limit and plasticity. The line on the chart is the classification boundary between the two soil types.

Another water-related property is the relative ease with which water flows through, or can be extracted from, the soil mass. Coarse-grained soils tend to be rapid draining, or permeable. Fine-grained soils tend to be nondraining, or impervious, and may literally seal out flowing water.

Soil structure may be classified in many ways. A major distinction is that made between soils that are considered to be *cohesive* and those considered *cohesionless*. Cohesionless soils are those consisting pre-

TABLE 2.1. Atterberg Limits for Water Content in Fine-Grained Soils

Description of Structural Character of the Soil Mass	Analagous Material and Behavior	Water Content Limit
Liquid	Thick soup; flows or is very easily deformed	
		Liquid limit: w_L
Plastic	Thick frosting or toothpaste; retains shape, but is easily deformed without cracking	Magnitude of range is *Plasticity Index:* I_p
		Plastic limit: w_P
Semisolid	Cheddar cheese or hard caramel candy; takes permanent deformation but cracks	
		Shrinkage limit: w_S (Least volume attained upon drying out)
Solid	Hard cookie; crumbles up if deformed	

dominantly of sand and gravel with no significant bonding of the discrete soil particles. The addition of a small amount of fine-grained material will cause the cohesionless soil to form a weakly bonded mass when dry, but the bonding will virtually disappear with a small percentage of moisture. As the percentage of fine materials is increased, the soil mass becomes progressively more cohesive, tending to retain some defined shape right up to the fully saturated, liquid consistency.

The extreme cases of cohesive and cohesionless soils are typically personified by a pure clay and a pure, or clean, sand respectively. Typical soil mixtures will range between these two extremes, so they are useful in establishing the boundaries for classification. For a clean sand the structural nature of the soil mass will be largely determined by three properties: the particle shape (well rounded versus angular),

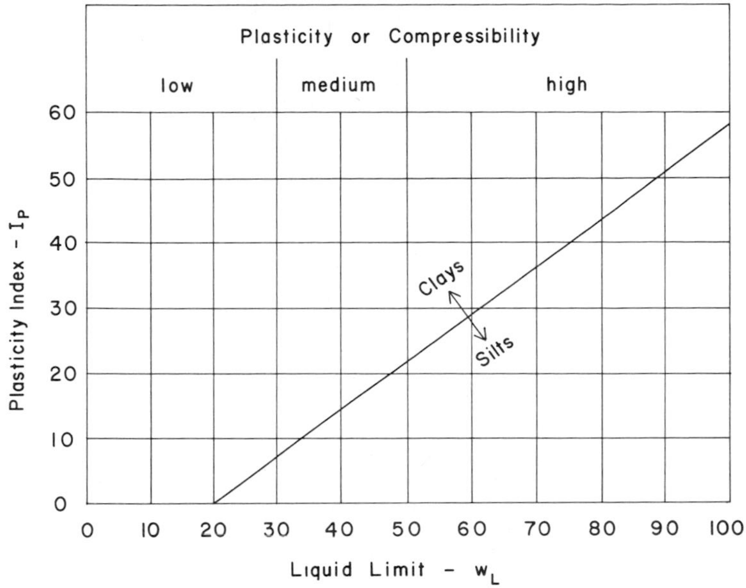

FIGURE 2.3. Plasticity chart–using Atterberg limits.

the nature of size gradation (well-graded, gap-graded or uniform), and the density or degree of compaction of the soil mass.

The density of a sand deposit is related to how closely the particles are fit together and is essentially measured by the void ratio. The actions of water, vibration and shock, and compressive force will tend to pack the particles into tighter arrangements. Thus the same sand particles may produce strikingly different soil deposits as a result of density variation.

Table 2.2 gives the range of density classifications that are commonly used in describing sand deposits, varying from very loose to very dense. The general character of the deposit and the typical range of usable bearing strength are shown as they relate to the density. As mentioned previously, however, the effective nature of the soil depends on additional considerations, principally the particle shape and the size gradation. Also of concern are the absolute particle size, generally established by the D_{10} property, and the amount of water present, as measured by w or S. A minor amount of water will often tend to give a slight cohe-

TABLE 2.2. Average Properties of Cohesionless Soils

Relative Density	Blow Count, N (blows/ft)	Void Ratio, e	Simple Field Test with $\frac{1}{2}$ in. Diameter Rod	Usable Bearing Strength (k/ft^2)
Loose	<10	0.65–0.85	Easily pushed in by hand	0–1.0
Medium	10–30	0.35–0.65	Easily driven in by hammer	1.0–2.0
Dense	30–50	0.25–0.50	Driven in by repeated hammer blows	1.5–3.0
Very dense	>50	0.20–0.35	Barely penetrated by repeated hammer blows	2.5–4.0

siveness to the sand, as the surface tension in the water partially bonds the discrete sand particles. When fully saturated, however, the sand particles are subject to a buoyancy effect that can work to substantially reduce the stability of the soil.

Many physical and chemical properties affect the structural character of clays. Major considerations are the particle size, the particle shape, and whether the particles are inorganic or organic. The percentage of water in a clay has a very significant effect on its structural nature, changing it from a rocklike material when dry to a viscous fluid when saturated. The property of a clay corresponding to the density of sand is its consistency, varying from very soft to very hard. The general nature of clays and their typical usable bearing strengths as they relate to consistency are shown in Table 2.3.

Another major structural property of fine-grained soils is relative plasticity. This was discussed in terms of the Atterberg limits and the classification was made using the plasticity chart shown in Figure 2.3. Most fine-grained soils contain both silt and clay, and the predominant character of the soil is evaluated in terms of various measured properties, most significant of which is the plasticity index. Thus identification as "silty" usually indicates a lack of plasticity (crumbly, friable,

TABLE 2.3. Average Properties of Cohesive Soils

Consistency	Unconfined Compressive Strength (k/ft^2)	Simple Field Test by Handling of an Undisturbed Sample	Usable Bearing Strength (k/ft^2)
Very soft	<0.5	Oozes between fingers when squeezed	0
Soft	0.5–1.0	Easily molded by fingers	0.5–1.0
Medium	1.0–2.0	Molded by moderately hard squeezing	1.0–1.5
Stiff	2.0–3.0	Barely molded by strong squeezing	1.0–2.0
Very stiff	3.0–4.0	Barely dented by very hard squeezing	1.5–3.0
Hard	4.0 or more	Dented only with a sharp instrument	3.0 +

etc.) while that of "claylike" or "clayey" usually indicates some significant degree of plasticity (moldable even when only partly wet).

Various special soil structures are formed by actions that help produce the original soil deposit or work on the deposit after it is in place. Coarse-grained soils with a small percentage of fine-grained material may develop arched arrangements of the cemented coarse particles resulting in a soil structure that is called *honeycombed*. Organic decomposition, electrolytic action, or other factors can cause soils consisting of mixtures of bulky and flaky particles to form highly voided soils that are called *flocculent*. The nature of formation of these soils is shown in Figures 2.4 and 2.5. Water deposited silts and sands, such as

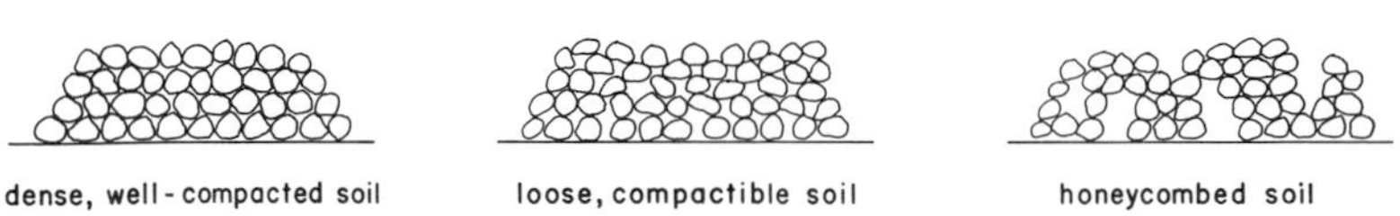

FIGURE 2.4. Soil structures in cohesionless soils.

FIGURE 2.5. **Soil structures in mixed-grain soils.**

those found at the bottom of dry streams or ponds, should be suspected of this condition if the tested void ratio is found to be quite high.

Honeycombed and flocculent soils may have considerable static strength and be quite adequate for foundation purposes as long as no unstabilizing effects are anticipated. A sudden, unnatural increase in the water content or significant vibration or shock may disturb the fragile bonding, however, resulting in major consolidation of the soil. This can produce major settlement of ground surfaces or foundations if the affected soil mass is large.

Behavior under stress is usually quite different for the two basic soil types: sand and clay. Sand has little resistance to stress unless it is confined. Consider the difference in behavior of a handful of dry sand and sand rammed into a strong container. Clay, on the other hand, has resistance to tension in its natural state all the way up to its liquid consistency. If a hard, dry clay is pulverized, however, it becomes similar to loose sand until some water is added.

In summary, the basic nature of structural behavior and the significant properties that affect it for the two soil types are as follows:

Sand. Little compression resistance without some confinement; principal stress mechanism is shear resistance (interlocking particles grinding together); important properties are angle of internal friction (ϕ), penetration resistance (N) to a driven object such as a soil sampler, unit density in terms of weight or void ratio, grain shape, predominant grain size and nature of size gradation. Some reduction in capacity with high water content.

Clay. Principal stress resistance in tension; confinement generally of concern only in soft, wet clays (to prevent flowing or oozing of the mass); important properties are the unconfined compressive strength

(q_u), liquid limit (w_L), plastic index (I_p), and relative consistency (soft to hard).

We must remind the reader that these represent the cases for pure clay and clean sand, which generally represent the outer limits for the range of soil types. Soil deposits typically contain some percentage of all three basic soil ingredients: sand, silt, and clay. Thus most soils are neither totally cohesive nor totally cohesionless and possess some of the characteristics of both of the basic extremes.

Soil classification or identification must deal with a number of properties for precise categorization of a particular soil sample. Many systems exist and are used by various groups with different concerns. The three most widely used systems are the triangular textural system used by the U.S. Department of Agriculture; the AASHO system, named for its developer, the American Association of State Highway Officials; and the so-called unified system, which is primarily used in foundation engineering. Each of these systems reflects some of the primary concerns of the developers of the system.

The unified system relates to properties of major concern in stress and deformation behavior, excavation and dewatering problems, stability under load, and other issues of concern to foundation designers. The triangular textural system relates to problems of erosion, water retention, ease of cultivation, and so on. The AASHO system relates primarily to the effectiveness of soils for use as base materials for pavements, both as natural soil deposits and as fill material. While the unified system is of major interest in foundation design, there is some value in being familiar with the other systems. A major reason is that in some situations information about sites may be available from highway or agricultural agencies and may be found useful for preliminary analysis and design or for more intelligently determining the nature and extent of soil exploration required for the site. Thus the ability to make some translations from one system to the other is useful.

Figure 2.6 shows the triangular textural system, which is given in graphic form and permits easy identification of the limits used to distinguish the named soil types. The property used is strictly grain size percentage, which makes the identification quite approximate since there are potential overlaps between very fine sand and silt and

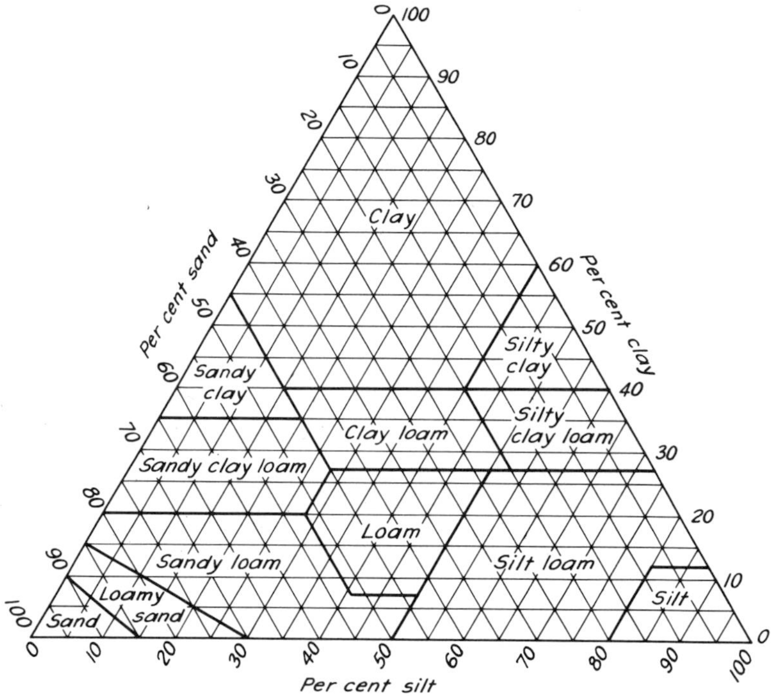

FIGURE 2.6. Triangular textural classification chart used by the U.S. Department of Agriculture. Reprinted from *Foundation Engineering* (Ref. 1) with permission of the publisher, John Wiley & Sons, Inc., New York.

between silt and clay. For agricultural concerns this is of less importance than it may be in foundation engineering. As important as grain size is in broadly characterizing the soil type, there are many more properties that the foundation designer must know, such as grain shape, size gradation, water content, plasticity, and so on.

Use of the textural graph consists of finding the percentages of the three basic soil types and projecting these three points on the edge of the graph to an intersection point that falls in one of the named groups. If the point is near the edge of the group, the soil will have some shared characteristics with the adjacent group. For example, a soil determined to have 46% sand (possibly including some gravel), 21% silt, and 33%

clay would fall in the group called "sandy clay loam." However, it would be close to the border of "clay loam" and would be somewhere between the two soil types in actual nature. If a foundation designer were to be told that the soil was somewhere between these two soil types, he could place the location approximately on the triangular graph and project outward to the edges to approximate the percentage of the sand, silt, and clay. That information could be extrapolated to predict a bracketed range of expectations for the behavior of the soil as a foundation material and also to anticipate what additional information would be particularly desirable from a soils investigation.

One useful purpose of the triangular graph is to observe the extent to which percentages of the various materials affect the essential nature of the soil. A sand, for example, must be relatively clean (free of fines) to be considered as such. With as much as 60% sand and only 40% clay the soil is considered essentially a clay. The general situation is that the finer the particles, the greater their influence on establishing the basic nature of the soil mass on a textural basis.

The AASHO system is shown in Table 2.4. The three basic items of data used are the grain size analysis, the liquid limit, and the plasticity index, the latter two properties relating only to fine-grained soils. On the basis of this data the soil type is located by group, and its general usefulness as a base for paving is rated, ranging generally from excellent for sand and gravel to poor for clay. This system is also of limited use for foundation design, although it is somewhat more informative than the textural system, especially for fine-grained materials. The AASHO system is also quite useful in evaluating the effectiveness of site materials as bases for floor slabs on grade, for sidewalks, and for other site paving. As with the textural system, structural properties can only be broadly established and the principal worth is in giving a head start to more extensive soil investigation.

The unified system is shown in Figure 2.7. It consists of categorizing the soil into one of 15 groups, each identified by a two-letter symbol. As with the AASHO system, the primary data used are the grain size analysis, the liquid limit, and the plasticity index. It is thus not significantly superior to that system in terms of its data base, but it provides more distinct identification of the soil pertaining to significant considerations of structural behavior.

None of these systems provides information sufficient for founda-

TABLE 2.4. American Association of State Highway Officials Classification of Soils and Soil Aggregate Mixtures—AASHO Designation M-145

General Classification [a]	Granular Materials (35 per cent or Less Passing No. 200)							Silt-Clay Materials (More than 35 per cent Passing No. 200)			
Group Classification	*A*–1		*A*–3	*A*–2				*A*–4	*A*–5	*A*–6	*A*–7
	A–1–*a*	*A*–1–*b*		*A*–2–4	*A*–2–5	*A*–2–6	*A*–2–7				*A*–7–5, *A*–7–6
Sieve analysis per cent passing:											
No. 10	50 max										
No. 40	30 max	50 max	51 min								
No. 200	15 max	25 max	10 max	35 max	35 max	35 max	35 max	36 min	36 min	36 min	36 min
Characteristics of fraction passing No. 40:											
Liquid limit				40 max	41 min	40 max	41 min	40 max	41 min	40 max	41 min
Plasticity index	6 max		N.P. [b]	10 max	10 max	11 min	11 min	10 max	10 max	11 min	11 min
Usual types of significant constituent materials	Stone fragments—gravel and sand		Fine sand	Silty or clayey gravel and sand				Silty soils		Clayey soils	
General rating as subgrade	Excellent to good							Fair to poor			

[a] Classification procedure: With required test data in mind, proceed from left to right in chart; correct group will be found by process of elimination. The first group from the left consistent with the test data is the correct classification. The *A*–7 group is subdivided into *A*–7–5 or *A*–7–6 depending on the plastic limit. For $w_P < 30$, the classification is *A*–7–6; for $w_P \geqq 30$, *A*–7–5.
[b] N.P. denotes nonplastic.

Major Divisions			Group Symbols	Typical Names		Classification Criteria	
Coarse-Grained Soils More than 50% retained on No. 200 sieve	Gravels 50% or more of coarse fraction retained on No. 4 sieve	Clean Gravels	*GW*	Well-graded gravels and gravel-sand mixtures, little or no fines	Classification on basis of percentage of fines: Less than 5% Pass No. 200 sieve *GW, GP, SW, SP*; More than 12% Pass No. 200 sieve *GM, GC, SM, SC*; 5% to 12% Pass No. 200 sieve Borderline classification requiring use of dual symbols	$C_u = D_{60}/D_{10}$ Greater than 4; $C_z = \frac{(D_{30})^2}{D_{10} \times D_{60}}$ Between 1 and 3	
			GP	Poorly graded gravels and gravel-sand mixtures, little or no fines		Not meeting both criteria for *GW*	
		Gravels with Fines	*GM*	Silty gravels, gravel-sand-silt mixtures		Atterberg limits plot below "*A*" line or plasticity index less than 4	Atterberg limits plotting in hatched area are borderline classifications requiring use of dual symbols
			GC	Clayey gravels, gravel-sand-clay mixtures		Atterberg limits plot above "*A*" line and plasticity index greater than 7	
	Sands More than 50% of coarse fraction passes No. 4 sieve	Clean Sands	*SW*	Well-graded sands and gravelly sands, little or no fines		$C_u = D_{60}/D_{10}$ Greater than 6; $C_z = \frac{(D_{30})^2}{D_{10} \times D_{60}}$ Between 1 and 3	
			SP	Poorly graded sands and gravelly sands, little or no fines		Not meeting both criteria for *SW*	
		Sands with Fines	*SM*	Silty sands, sand-silt mixtures		Atterberg limits plot below "*A*" line or plasticity index less than 4	Atterberg limits plotting in hatched area are borderline classifications requiring use of dual symbols
			SC	Clayey sands, sand-clay mixtures		Atterberg limits plot above "*A*" line and plasticity index greater than 7	
Fine-Grained Soils 50% or more passes No. 200 sieve	Silts and Clays Liquid limit 50% or less		*ML*	Inorganic silts, very fine sands, rock flour, silty or clayey fine sands		(see plasticity chart)	
			CL	Inorganic clays of low to medium plasticity, gravelly clays, sandy clays, silty clays, lean clays			
			OL	Organic silts and organic silty clays of low plasticity			
	Silts and Clays Liquid limit greater than 50%		*MH*	Inorganic silts, micaceous or diatomaceous fine sands or silts, elastic silts			
			CH	Inorganic clays of high plasticity, fat clays			
			OH	Organic clays of medium to high plasticity			
Highly organic soils			*Pt*	Peat, muck and other highly organic soils		Visual-manual identification	

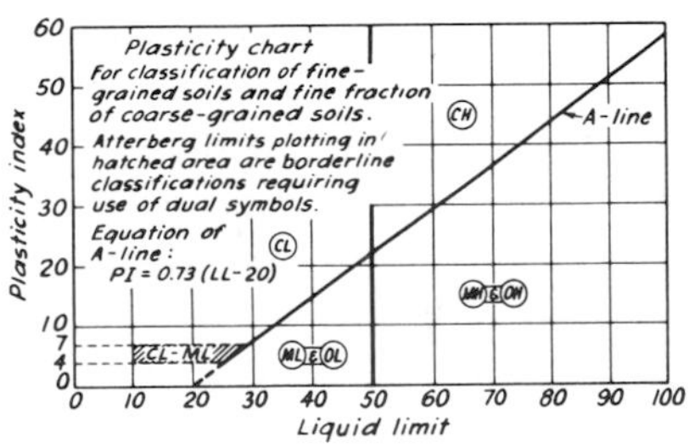

FIGURE 2.7. Unified system classification of soils for engineering purposes (ASTM Designation D-2487). Reprinted from *Foundation Engineering* (Ref. 1) with permission of the publisher, John Wiley & Sons, Inc., New York.

tion design, except for very conservative approximations. What is mainly lacking is any direct testing of the structural properties of the soil (such as penetration resistance of sand or unconfined compression strength of clay), especially in its undisturbed condition at the site. The information obtained from the classification system must therefore be added to that obtained from site observations, site tests, and lab tests for a complete set of data useful for good engineering design.

Building codes and engineering handbooks often use some simplified system of grouping soil types for the purpose of regulating or recommending foundation design criteria and some construction details. This topic is discussed in Section 2.6 with regard to the establishment of foundation design criteria.

2.3 Behavior of Foundations

Foundations are essentially elements that affect a transition between the building and the ground. There are thus three general areas of concern in foundation design.

1. The nature of the structure that must be supported; its size, shape, weight, type of structural system, sensitivity to deformation, and so on.
2. The nature of the ground, particularly the subsurface materials directly involved in resisting the foundation loads.
3. The structural actions of foundation elements, involving internal stresses and strains and the means by which they achieve the transfer of loads to the ground.

Most foundations consist of some elements of concrete, primarily because of the relative cost of the material and its high resistance to water, rot, insects, and various effects resulting from being buried in the ground. The two fundamental types of foundations are shallow bearing foundations and deep foundations. This distinction has to do primarily with where the load transfer to the ground occurs. With shallow foundations it occurs near the bottom of the building; with deep foundations the load transfer involves soil strata at some distance below the building.

The most common types of shallow foundations are wall and column footings, consisting of concrete strips and pads poured directly on the ground and directly supporting structural elements of the building. The basic stress transfer between the footing and the ground is by direct contact bearing pressure, inducing general mechanisms of soil behavior illustrated in Figures 2.8, 2.9, and 2.10. Occasionally several structural elements of the building may be supported by a single large footing. The ultimate extension of this is to turn the entire underside of the building into one large footing, simulating the action of the hull of a ship. This is actually done in rare cases and, indeed, such a foundation does literally float on the soil and is called a raft.

If the soil at the bottom of the building is not adequate for the load transfers, or possibly is underlain by weak materials, it becomes necessary to utilize the resistance of lower soil strata. This may require going all the way down to bedrock, or merely to some more desirable soil layer. The technique for accomplishing this is simply to place the building on stilts, or tall legs, in the ground. The two basic types of elements used to do this are piles and piers. Piles are elements that are driven into the ground, much the same as a nail is driven into wood. Piers are shafts that are excavated and then filled with concrete.

Both piles and piers may be extended to establish a bearing at their ends, in rock or simply a very hard soil stratum. When not bearing on rock, piers are usually widened, or flared, at their ends to increase the contact bearing area. Piles may also be simply driven until their surfaces develop sufficient skin friction with the soil (as the nail does with the wood) when rock or really hard soil strata are not within reach.

With some soils or water conditions it is sometimes necessary to line the pier excavation until the concrete is deposited. If the shaft diameter is large, it may be necessary to use the caisson method, similar to that used for the construction of bridge piers below the water level. Although this is not often necessary, piers are still commonly referred to as caissons.

The problems of behavior and design of shallow bearing foundations are discussed in Chapter 3. The design of typical elements is illustrated, and tables of predesigned, typical elements are given. The design of deep foundations is generally beyond the scope of simplified engineering techniques. Some of the general principles of their behavior, examples of typical elements used, and various problems encountered in design and construction of buildings on deep foundations are discussed and illustrated in Chapter 4.

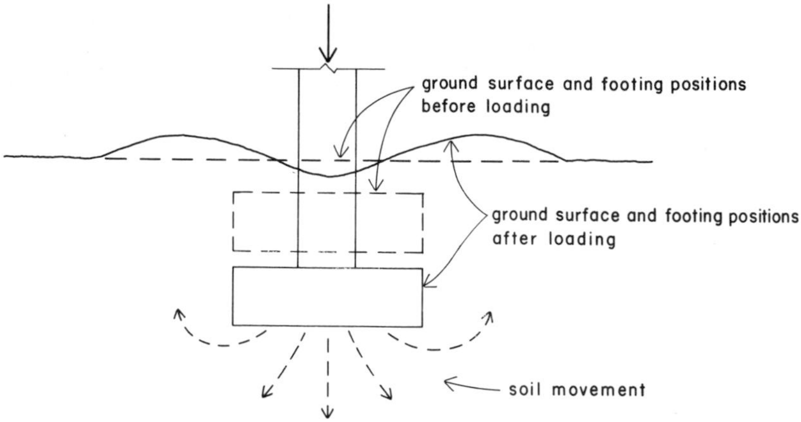

FIGURE 2.8. Typical failure mechanism for a simple bearing footing.

When loads are applied to a bearing foundation, stresses are generated in the soil mass. In order to visualize these stresses and the accompanying strains it is necessary to consider the nature of movement of the foundation and the soil mass. Figure 2.8 shows the typical failure mechanism for a simple bearing footing as it is pushed into a soil mass. Part of the vertical movement of the footing is accounted for by the consolidation, or squeezing, of the soil immediately beneath the footing. If any additional movement of the footing is to occur, it must be accomplished by pushing some of the soil out from under the footing, which then involves stresses and movements in the soil mass adjacent to and even above the footing.

Figure 2.9 shows the so-called bulbs of pressure that occur in a typical compressible soil beneath a bearing footing. The contour lines of pressure indicate both the net direction of the pressure and the location of equal points of pressure magnitude in terms of percentages of the contact pressure, q, at the bottom of the footing. Although the foundation load is directed vertically downward, the net pressure is vertically downward only in the soil mass immediately beneath the center of the footing. As we move away from the center, the net pressure becomes increasingly horizontal. Adjacent to the footing and above the level of its bottom the net pressure will actually be upward, if the footing is pushed into the soil mass.

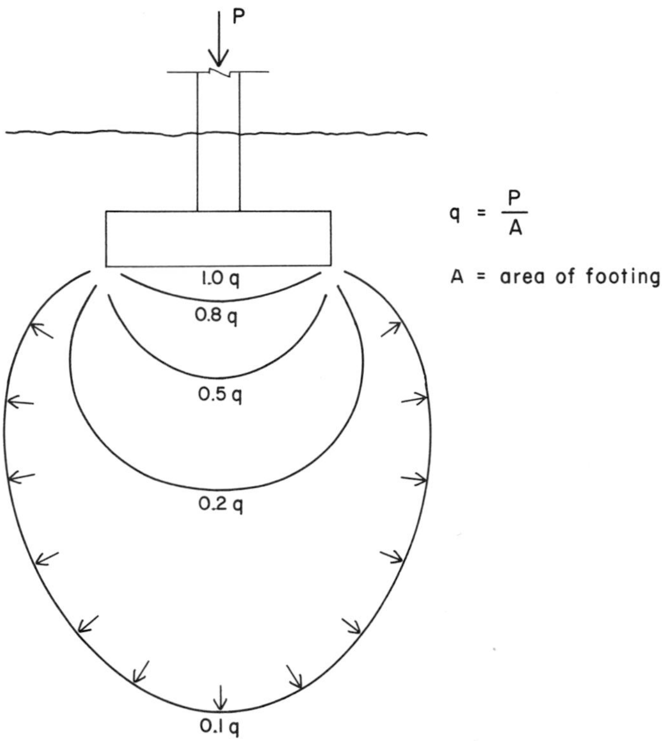

FIGURE 2.9. Bulbs of equal pressure under a bearing footing.

In order to visualize this total effect we consider three zones of the soil mass, as shown in Figure 2.10. In Zone A, immediately beneath the footing, the soil is squeezed biaxially or triaxially by the vertical forces of the footing (F_2) opposed by the mass of soil below Zone A (F_3) and by the confining restraint provided by the adjacent soil masses (F_4). The F_4 force on Zone B produces the three reactive forces, F_5, F_6, and F_7. F_5 and F_6 are produced by the soil masses adjacent to and below Zone B. F_7 is essentially due to the weight of the soil in Zone C, indicated as F_8. If there is additional weight of elements on top of the ground surface, such as paving slabs, it may be added to F_8.

The collective effect of all of these forces on the footing depends on various properties of the soil materials in the various zones. F_3 must be

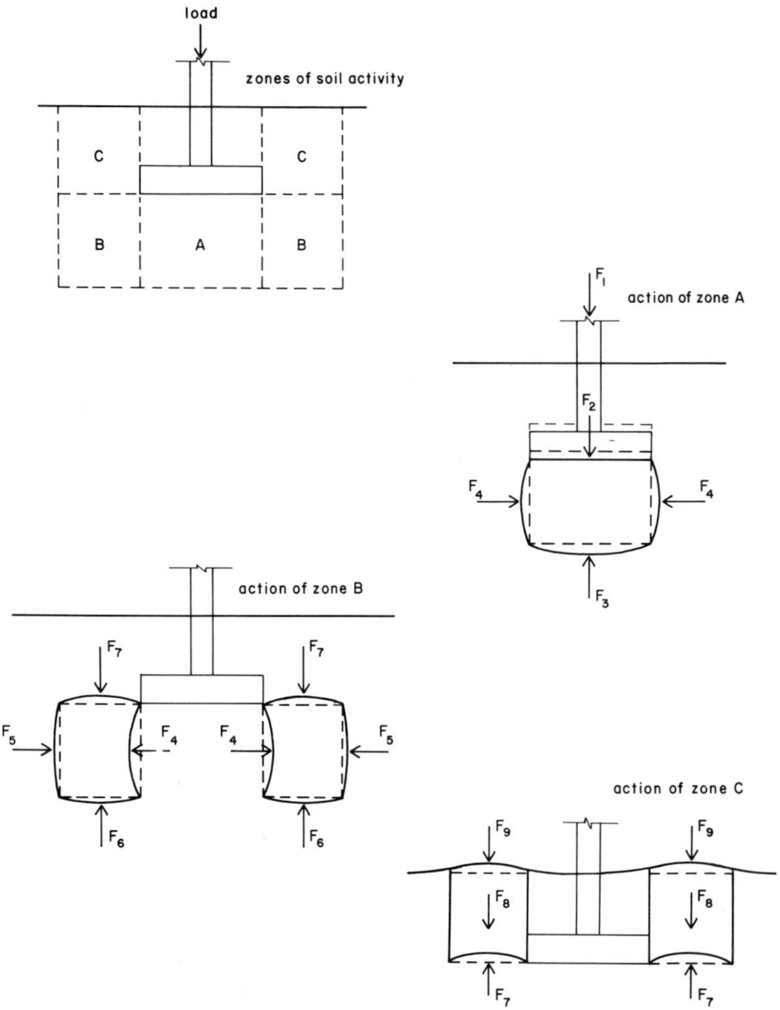

FIGURE 2.10. Zones of soil activity for a bearing footing.

generated by the soil mass beneath Zone A, and if this soil is highly compressible, there may be considerable settlement, even though the soil in Zone A is quite strong. If the footing is very wide, the F_3 pressure may be significant to some depth. Figure 2.11 shows a typical subsurface profile with stratified layers of soils with different properties. If the footings placed at the same level are considerably different in width, there will be significant pressure at some depth below the wider footings. Thus the existence of the highly compressible soil in Stratum 4 will have a negligible effect on the narrower footings, but may produce some significant settlement of the wider footings.

The relative importance of F_4 is somewhat different if the soil in Zone A is essentially cohesive as opposed to cohesionless. In a cohesionless soil F_4 is absolutely mandatory if there is to be any stress resistance by the soil mass in Zone A. Sand cannot develop significant resistance to vertical force unless it has the containing force represented by F_4. In a predominantly clay soil, on the other hand, as long as the clay is not in a liquid or highly plastic state, the tensile strength of the soil permits some development of vertical stress without any confining force. If the clay is at all soft, however, the absence of an adequate F_4 force will cause the soil mass to ooze horizontally under vertical stress.

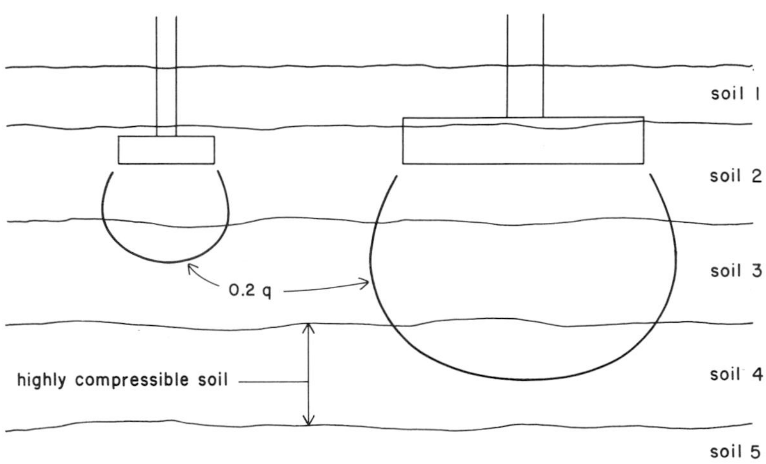

FIGURE 2.11. Difference in pressure effects under footings of different width.

This is simply a matter of the liquidlike material moving in the direction of the least pressure resistance upon being squeezed.

The type of soil in Zone B may also affect the relative importance of the restraining forces on this mass. If the soil is cohesive, Zone B can function with the F_4 force opposed only by the F_5 constraint. If the soil in Zone B is cohesionless, however, the effect of force F_7 becomes more critical in increasing the degree of contraint for both zones B and A. Thus the distance of the bottom of the footing from the surrounding ground surface is of considerable significance in cohesionless soils.

All of the actions described relate to the determination of two primary factors: the permissible contact bearing pressure at the bottom of the footing and the anticipated total vertical settlement of the footing. As illustrated in Figures 2.9 and 2.10 the contact bearing pressure is determined from the force F_2 as

$$q = \frac{F_2}{A}$$

where A is the area of the bottom of the footing.

The limiting value for F_2 is not determined solely by the properties of the soil mass immediately beneath the footing. The limit for F_2, or for the permissible soil bearing pressure, q, is developed from consideration of all of the force actions shown in the illustrations in Figure 2.10, with major concern for F_3, F_4, and F_7. Thus the general considerations for establishment of the allowable loads on bearing footings may be summarized as follows:

1. The vertical compressive stress resistance and deformation resistance of the soil in Zone A.
2. The vertical compressive stress resistance and deformation resistance of soil masses below Zone A if the footing is of significant width.
3. The lateral (horizontal) pressure resistance of the soil in Zone B.
4. The weight of the soil mass in Zone C, as a product of the soil density and the depth of the footing below the ground surface.
5. The width of the footing, as it influences the depth to which soil pressure must be considered in the supporting soil mass and the possibility of differential settlement in a series of footings of varying width.

In specific situations any one of these factors may be critical in the establishment of the limiting bearing stress. Even though the soil in Zone A is excellent for bearing, the existence of very weak strata beneath Zone A may allow only a fraction of the Zone A soil's capacity to be utilized, especially for very wide footings. If the soil in Zone A is of a type that is highly dependent on confinement, such as a clean sand or soft clay, reduction of the resistance from Zone B may be critical. Loss of this confinement can occur because of adjacent excavation or in steep hillside conditions. If the footing is placed only a short distance below ground, the resulting lack of confinement for Zone B may be critical. Finally, the soil in Zone A may be too weak, too compressible, of questionable stability, or highly sensitive to disturbances during the excavation and construction work.

There are of course other foundation-ground force interactions of concern besides the primary one of vertical load transfer. Many additional stress conditions and force interactions are discussed in other portions of this book. Some of those of major concern are the following:

Resistance to horizontal forces. Forces due to wind, earthquakes, horizontal earth pressures, and horizontal support forces from arches, rigid frames, and so on produce the need for horizontal as well as vertical resistance by the supporting ground. This problem is discussed in general in Chapter 5.

Moment-resistive footings. Freestanding walls, retaining walls, towers and chimneys represent structures that typically require foundations with resistance to overturning moments as well as the usual resistance to direct vertical and horizontal forces. The general problems of moment-resistive footings are discussed in Section 3.11.

Interrelated behavior of adjacent foundation elements. Various problems may occur when independent foundation elements are supported by the same ground mass. Some examples are the following:

Closely spaced footings. Closely spaced footings may produce undesirable overlaps of pressure in the soil or may simply make construction of separate foundations difficult. This issue is discussed in Section 3.8 with regard to closely spaced columns in buildings.

Closely spaced deep foundation elements. These may also be a source of construction problems, but have some special stress problems as well.

Vertically separated adjacent footings. Closely spaced bearing footings occurring at significantly different elevations can generate some special problems. This problem is discussed in general in Section 3.15.

Load sharing in foundation systems. Various situations of this kind occur, sometimes involving stress sharing in the soil and other times involving development of foundation elements to achieve load sharing or load distribution. Some examples are:

Tying of isolated footings for sharing lateral loads. This is a common technique for transferring lateral loads from less resistive, or heavier loaded, foundations to other elements in the system. This issue is discussed in Section 3.15 for shallow bearing foundations.

Use of foundation walls as distributing elements. Where they exist, foundation walls are often used as ties, spanning elements, bracing elements, and so on. Some of these functions are discussed in general in Section 3.1. The special case of columns existing in walls is discussed in Section 3.10.

2.4 Special Soil Problems

The brief discussion of soil properties and stress behaviors that is given in this book is generally sufficient for many relatively ordinary soil conditions. A great number of special soil situations can be of major concern in foundation design. Some of these are predictable on the basis of regional climate and geological situations. Anyone expecting to do a major amount of foundation design work should study soil mechanics in much greater depth than it is presented in this work. A few of the special problems of particular concern are discussed in the following material.

Expansive Soils. In climates with long dry periods, fine-grained soils often shrink to a minimum volume, sometimes producing vertical cracking in the soil masses that extends to considerable depths. When significant rainfall occurs, two phenomena occur that can produce problems for structures. The first is the swelling of the ground mass as water is absorbed, which can produce considerable upward or sideways pressures on structures. The second is the rapid seepage of water into lower soil strata through the vertical cracks.

The soil swelling can produce major stresses in foundations, especially when it occurs nonuniformly, which is the general case because of paving, landscaping, and so on. Compensation for these stresses depends on the details of the building construction, the type of foundation system, and the relative degree of expansive character in the soil. Local building codes usually have provisions for design with these soils in regions where they are common. If this property is to be expected, it is highly advised that the tests necessary to establish the expansive property be performed and the results together with necessary considerations for foundation design be reviewed by an experienced foundation engineer.

Collapsing Soils. Collapsing soils are in general soils with large voids. The collapse mechanism is essentially one of rapid consolidation as whatever tends to maintain the soil structure in the large void condition is removed or altered. Very loose sands may display such behavior when they experience drastic changes in water content or are subjected to shock or vibration. The more common cases, however, are those involving soil structures in which fine-grained materials achieve a bonding or molding of cellular voids. These soil structures may be relatively strong when dry but may literally dissolve when the water content is significantly raised. The bonded structures may also be destroyed by shock or simply by excessive compression stress.

This behavior is generally limited to a few types of soil and can usually be anticipated when such soils display large void ratios. Again, the phenomenon of collapse is often a local condition and is given special consideration in local building codes and practices. The two ordinary methods of dealing with collapsing soils are to stabilize the soil by introducing materials to partly fill the void and reduce the potential degree of collapse, or to use vibration, saturation, or other means to cause the collapse to occur prior to construction. When considerable site grading is to be done, it is sometimes possible to temporarily place soil to some depth on the site, providing sufficient compression to cause significant deformation of the potential foundation-bearing materials. The latter method is effective only for soils in which the static pressure thus produced is truly capable of causing significant consolidation.

Differential Settlements. It is generally desirable that the foundation of any building settles uniformly. If separate elements of the foundation settle significantly different amounts, there will be some distortion of the building structure. The seriousness of this situation depends upon the materials and type of the construction; the most critical cases are ones involving masonry, concrete and plaster constructions that tend to be quite rigid and subject to brittle cracking.

A number of situations can result in differential settlements. Some of these can be adjusted for by careful design of the foundation whereas others are more difficult to compensate for. The following are some of the situations that can cause this problem.

Nonuniform subgrade conditions. Pockets of poor soil and soil strata that are not horizontal can result in different settlement conditions for footings at different locations on the site. Any attempts to equalize settlements in this case require extensive information about the subsurface soil conditions at all points on the site where foundations occur. In addition, settlement calculations must be carefully done for each foundation, which can constitute considerable work in the structural design.

Footings of significantly different size. As discussed before, the vertical stresses under large footings can reach great depths. This can produce larger settlements of the large footings, even though the contact bearing pressure is the same for all footings. Precise compensation in design for this situation also requires extensive knowledge of subsurface conditions and entails laborious settlement calculations.

Footings placed at different elevations. Footings placed at different elevations may bear on different soil strata with significantly different settlement resistance, may have considerable difference in the containment due to overlying soil, or may result in some footings being below the water table and some above it. All of these conditions may result in the development of different settlements under the same soil pressure.

Varying ratios of live and dead loads. Although the foundations must carry all the building loads, the effect of dead load is often more critical. One reason for this is that the dead load tends to be more "real," while the live load is often quite vaguely established. Another reason is that the dead load is permanent and may thus have more influence on settlements that are progressive with time, such as those caused by clay soils or soils subject to repetitive effects such as fluctuations of water content or frost actions. Thus if footing bearing pressures are equal for the total load, the settlements due to dead load alone will be equal only when the percentage of dead load is the same on all footings. Design for equalized dead load pressures is discussed and illustrated in Section 3.15.

2.5 Soil Investigation

The amount of information necessary for a good foundation design depends on a number of factors. For a small building located on a flat site with good soil conditions the necessary information may be minimal. For a large building on a difficult site considerable site exploration and extensive field or laboratory testing may be necessary. In any case, the investigation of site and subsurface conditions and the interpretation of information obtained from them should be done by persons with experience and competence in this work.

It is not our purpose here to explain how to do such investigations, but rather to describe the usual processes, the form of information generally provided, and the relation in general of soil investigation to foundation design. We will also discuss the various types of information that may be useful and some of the possible sources of information other than formal soil testing service programs.

Visual inspection of the building site by the foundation designer is highly desirable. Although a site survey will give the actual ground surface contours and location of items such as existing constructions, streets, and so on, a visit to the site may provide much more information. Possible items of interest are:

Evidence of cut and fill. If there is evidence of considerable recontouring of the ground surface form, the original ground surface may be considerably below the present surface. For a building with heavy loads this may require extending the footings considerably below the surface to reach desirable bearing material.

Patterns of local fault. On the site and in the area around it there may be evidence of erosion, surface subsidence, land slippage on slopes, or other phenomena that indicate unstable local ground conditions.

Condition of existing structures. If there is any major construction on or near the site, it should be inspected for evidence of damaging settlements, effects of frost heave or expansive soils, and so on.

Persons with some experience in soil investigation may be able to determine some major aspects of the subsurface conditions without elaborate equipment or tests. Soil samples to some depth may be obtained with a posthole digger or a hand auger, and the general character as well as some fairly significant properties may be reasonably determined from handling these samples. Color, odor, general texture, moisture content, density, and ease of excavation of the samples can be correlated to give a fair approximation of the classification and average structural properties of the soil. This type of investigation may be quite adequate for small projects where no unusual conditions exist. It must be emphasized that while the investigation may be simple, the person performing it should have considerable knowledge about soils and their structural properties, and in addition should preferably be familiar with the local climate conditions and regional geology as well as foundation design and construction practices.

A soil investigation performed by a soil testing service can be quite expensive if explorations must be made to a great depth and at several locations on the site. Before such an investigation is performed it is highly desirable that there be some information about the building size, location, and type of construction, and preferably some idea about the general subsurface conditions that can be anticipated. In such a case the desired location for the soil borings, the necessary depth to which they must be carried, and the extent of testing of samples required can be more intelligently planned. If soil exploration is done on a large site with virtually unknown subsurface conditions and little idea about the location or type of construction that is planned, a second exploration may be necessary. This may not be a serious economic problem with a large building project, but on a small project can result in the total soil investigation cost being as much as the engineering fee for the entire foundation design. It is thus desirable to explore all possible sources of information prior to ordering formal testing by a professional organization.

Depending on the location of the site, there may be a lot of opportunity to obtain preliminary information about the ground conditions. Various potential sources are the following:

Site visit. As previously described, a visit to the site can be quite informative. This is best done by the foundation designer, but when that is not possible may be done by others with some instruction from the designer about what to look for.

Simple site exploration. Exploration at the site is also best done by someone with experience in foundation design, but there is some value in information and samples obtained by others and sent to the designer for analysis.

Results from explorations on adjacent property or from previous construction on the site. Depending on the proximity to the currently planned work, these may be quite useful. In any event, they can help for preliminary planning.

Information from various organizations. Many people are interested in subsurface ground conditions. In many cases information may be obtained free or by paying a small charge for reproduction or handling. Some possible sources are:

City, county, or state engineering or building departments.
County and state highway departments.
County, state, and federal agricultural departments.
Agencies involved in studies for water resources, erosion control, geological surveys, and so on.
Army, Navy and Air Force studies for nonclassified projects.

Although information from these groups may not include all of the facts desirable for foundation design, it can still be useful for preliminary design and feasibility studies.

Actually, soil testing organizations usually avail themselves of any such information that is available for the region in which they work. Thus even before contracting to perform a formal testing program, they will often supply to designers some information about the general conditions that can be anticipated for a particular site.

A formal soil testing program ordinarily consists of the following:

Exploration of the subsurface conditions by some means to obtain samples of soil at various levels below the ground surface.

Field tests consisting of some observations made during the exploration as well as observations and tests on the samples obtained.

Laboratory tests on some samples, such as determination of dry weight, particle size gradation, and so on.

Interpretation of the information obtained and recommendations for criteria and some details of the foundation design.

The extent of exploration, the type of equipment used, the type of tests performed, and the soil properties determined are all subject to considerable variation and require some judgment on the part of those conducting the investigation. The actual type of soil encountered has a lot to do with this. In many cases the information recorded and the terminology of identification of samples is done in relation to local practices and the requirements of local codes and building regulatory agencies.

When the need for deep foundations is anticipated, the exploration will generally be carried to a significant depth. If unusual soil conditions are discovered, more extensive testing may be performed. If separate explorations indicate nonuniform soil strata, many explorations may be necessary in order to obtain a clear picture of the contour of subsurface soil strata. On the other hand, if building loads are modest and the soil encountered is of generally good character for bearing, the necessary exploration and testing may be quite minimal.

For foundation design purposes, soil properties may be broadly separated into two groups. The first consists of those properties that are significant to the identity of the soil type. With reference to classification by the unified system, these are the properties necessary to establish the soil identity as one of the 15 types in the unified system chart, or in some cases, as a soil with marginal properties between two closely related types. For sand, the significant properties are the amount of fines and the size gradation characteristics of the coarse fraction. For silt and clay, the presence of organic material, the liquid limit, and the plasticity index are the major factors affecting identity.

The second group of soil properties includes those that relate directly to the structural character or stress and deformation behavior of the soil. While some of these properties can be presumed in a general way on the basis of the soil identity, there are some specific tests that provide more information, permitting more accurate predictions of structural performance.

For a sand, there are typically four items of information not included in the data used for classification by the unified system that are significant to structural behavior. These are the following:

Grain shape. The shape ranges from well rounded to angular.

Water content. Water content is expressed as measured in the natural state, but also as anticipated on the basis of annual climate variations.

Density. Density ranges from loose to dense and indicates the potential degree of consolidation and the settlement that will result from it.

Penetration resistance. Penetration resistance is usually quoted as the N value, which is the number of blows required to advance a particular type of soil sampler into the soil deposit.

For a clay, the principal tested structural property is the unconfined compression strength, q_u. This may be approximated by some simple field tests, but is most accurately established by a laboratory test on a carefully excavated, so-called undisturbed sample of the soil.

Many soil properties are interrelated or derivative; thus a crosscheck is possible when considerable information is available. Thus unit dry weight and penetration resistance are both related to the relative density of sand. Similarly, unconfined compression strength, relative consistency, and plasticity are interrelated for clay.

The structural character of silts ranges considerably from that resembling a low-plasticity clay to that resembling a fine sand. Thus it is sometimes necessary to use tested properties pertinent to both cohesive and cohesionless soils for complete evaluation of the structural character of silty soil.

2.6 Foundation Design Criteria

For the design of ordinary bearing-type foundations several structural properties of a soil must be established. The principal values are the following:

Allowable bearing pressure. This is the maximum permissible value for vertical compression stress at the contact surface of bearing elements. It is typically quoted in units of pounds or kips per square foot of contact surface.

Compressibility. This is the predicted amount of volumetric consolidation that determines the amount of settlement of the foundation. Quantification is usually done in terms of the actual dimension of vertical settlement predicted for the foundation.

TABLE 2.5. Summary of Properties and Recommended Design Values for Typical Soils

Description	Gravel, well-graded; little or no fines		
ASTM classification (See Figure 2.7)	GW		
Significant properties	≧95% retained on No. 200 sieve (0.003 in.) ≧50% of coarse fraction retained on No. 4 sieve ($\frac{3}{16}$ in.) $C_u > 4$ $1 < C_z < 3$		
Field identification	Significant amounts of coarse rock fragments; easily pulverized; fast draining; wide range of grain sizes		
Average properties	Loose	Medium	Dense
Allowable bearing (lb/ft^2)	($N < 10$)	($10 < N < 30$)	($N > 30$)
with minimum of one ft surcharge	1300	1500	2000
increase for surcharge (%/ft)	20	20	20
maximum total	8000	8000	8000
Lateral pressure			
active coefficient	0.25	0.25	0.25
passive (lb/ft^2 per ft depth)	200	300	400
Friction (coefficient or lb/ft^2)	0.50	0.60	0.60
Weight (lb/ft^3)			
dry	100	110	115
saturated	125	130	135
Compressibility	Medium	Low	Very low

TABLE 2.5. *(Continued)*

Gravel, poorly graded, little or no fines			Silty gravel and gravel-sand-silt mixes		
GP			GM		
$\geqq$95% retained on No. 200 sieve (0.003 in.) $\geqq$50% of coarse fraction retained on No. 4 sieve ($\frac{3}{16}$ in.) Does not meet C_u and/or C_z requirements for well-graded (GW)			50–88% retained on No. 200 sieve (0.003 in.) $\geqq$50% of coarse fraction retained on No. 4 sieve ($\frac{3}{16}$ in.) Atterberg plot below A line or $I_p < 4$		
Significant amounts of coarse rock fragments; easily pulverized; fast draining; has narrow range of sizes or is gap-graded			Gravely but forms clumps that pulverize with moderate effort; wet sample takes little or no remolding before disintegrating; slow draining		
Loose ($N < 10$)	Medium ($10 < N < 30$)	Dense ($N > 30$)	Loose ($N < 10$)	Medium ($10 < N < 30$)	Dense ($N > 30$)
1300	1500	2000	1000	1500	2000
20	20	20	20	20	20
8000	8000	8000	8000	8000	8000
0.25	0.25	0.25	0.30	0.30	0.30
200	300	400	167	250	333
0.50	0.60	0.60	0.40	0.50	0.50
90	100	110	100	115	130
120	125	130	125	135	145
Medium	Low	Very low	Medium	Low	Very low

TABLE 2.5. *(Continued)*

Description	Clayey gravel and gravel-sand-clay mixes		
ASTM classification (See Figure 2.7)	GC		
Significant properties	50–88% retained on No. 200 sieve (0.003 in.) ≧50% of coarse fraction retained on No. 4 sieve ($\frac{3}{16}$ in.) Atterberg plot above A line or $I_p > 7$		
Field identification	Gravely but forms hard clumps that require considerable effort to pulverize; wet sample takes some remolding before disintegrating; very slow draining		
Average properties	Loose	Medium	Dense
Allowable bearing (lb/ft^2)	($N < 10$)	($10 < N < 30$)	($N < 30$)
with minimum of one ft surcharge	1000	1500	2000
increase for surcharge (%/ft)	20	20	20
maximum total	8000	8000	8000
Lateral pressure			
active coefficient	0.30	0.30	0.30
passive (lb/ft^2 per ft depth)	167	250	333
Friction (coefficient or lb/ft^2)	0.40	0.50	0.50
Weight (lb/ft^3)			
dry	110	120	125
saturated	125	130	135
Compressibility	Medium	Low	Very low

TABLE 2.5. *(Continued)*

Sand, well-graded; gravely sand; little or no fines	Sand, poorly graded; gravely sand; little or no fines
SW	SP
≧95% retained on No. 200 sieve (0.003 in.) >50% of coarse fraction passes No. 4 sieve ($\frac{3}{16}$ in.) $C_u > 6$ C_z 1–3	≧95% retained on No. 200 sieve (0.003 in.) >50% of coarse fraction passes No. 4 sieve ($\frac{3}{16}$ in.) Does not meet C_u and/or C_z requirements for well graded (SW)
Relatively clean sand with wide size range; easily pulverized; fast draining	Relatively clean sand with narrow size range or gaps in grading; easily pulverized; fast draining

Loose ($N < 10$)	Medium ($10 < N < 30$)	Dense ($N > 30$)
1000	1500	2000
20	20	20
6000	6000	6000
0.25	0.25	0.25
183	275	367
0.35	0.40	0.40
100	110	115
125	130	135
Medium high	Medium	Low

Loose ($N < 10$)	Medium ($10 < N < 30$)	Dense ($N > 30$)
1000	1500	2000
20	20	20
6000	6000	6000
0.25	0.25	0.25
75	150	200
0.35	0.40	0.40
90	100	110
120	125	130
Medium high	Medium	Low

TABLE 2.5. *(Continued)*

Description	Silty sand and sand-silt mixes		
ASTM classification (See Figure 2.7)	SM		
Significant properties	50–80% retained on No. 200 sieve (0.003 in.) >50% of coarse fraction passes No. 4 sieve ($\frac{3}{16}$ in.) Atterberg plot below A line, or $I_p < 4$		
Field identification	Sandy soil; forms clumps that can be pulverized with moderate effort; wet sample takes little remolding before disintegrating; slow draining		
Average properties	Loose	Medium	Dense
Allowable bearing (lb/ft^2)	($N < 10$)	($10 < N < 30$)	($N > 30$)
with minimum of one ft surcharge	500	1000	1500
increase for surcharge (%/ft)	20	20	20
maximum total	4000	4000	4000
Lateral pressure			
active coefficient	0.30	0.30	0.30
passive (lb/ft^2 per ft depth)	100	167	233
Friction (coefficient or lb/ft^2)	0.35	0.40	0.40
Weight (lb/ft^3)			
dry	105	115	120
saturated	125	130	135
Compressibility	Medium	Low	Low

TABLE 2.5. *(Continued)*

Clayey sand and sand-clay mixes			Inorganic silt, very find sand, rock flour, silty or clayey fine sand		
SC			ML		
50–80% retained on No. 200 sieve (0.003 in.) >50% passes No. 4 sieve ($\frac{3}{16}$ in.) Atterberg plot above A line or $I_p > 7$			$\geqq$50% passes No. 200 sieve (0.003 in.) $w_L \leqq 50\%$ Atterberg plot below A line $I_p < 20$		
Sandy soil; forms clumps that offer some resistance to being pulverized; wet sample takes some remolding before disintegrating; very slow draining			Fine-grained soils of low plasticity; slow draining; dry clumps easily pulverized; won't form thin thread when molded		
Loose or soft ($N < 10$)	Medium ($10 < N < 30$)	Dense or stiff ($N > 30$)	Loose or soft ($N < 10$)	Medium ($10 < N < 30$)	Dense or stiff ($N < 30$)
1000	1500	2000	500	750	1000
20	20	20	20	20	20
4000	4000	4000	3000	3000	3000
0.30	0.30	0.30	0.35	0.35	0.35
133	217	300	67	100	133
0.35	0.40	0.40	0.35 or 250	0.40 or 375	0.40 or 500
105	115	120	105	115	120
125	130	135	125	130	135
Medium	Low	Low	Medium high	Medium	Low

TABLE 2.5. *(Continued)*

Description	Lean clay; inorganic clay of low to medium plasticity; gravely clay; sandy clay; silty clay
ASTM classification (See Figure 2.7)	CL
Significant properties	$\geqq$50% passes No. 200 sieve (0.003 in.) $w_L \leqq 50\%$ Atterberg plot above A line $I_p > 7$
Field identification	Fine-grained soil of low plasticity; slow draining; dry clumps quite hard, but not very difficult to pulverize

Average properties	Soft	Medium	Stiff
Allowable bearing (lb/ft^2)			
with minimum of one ft surcharge	1000	1500	2000
increase for surcharge (%/ft)	20	20	20
maximum total	3000	3000	3000
Lateral pressure			
active coefficient	0.40	0.50	0.75
passive (lb/ft^2 per ft depth)	267	467	667
Friction (coefficient or lb/ft^2)	500	750	1000
Weight (lb/ft^2)			
dry	80	95	105
saturated	110	120	130
Compressibility	High	Medium high	Low

TABLE 2.5. *(Continued)*

Description	Fat clay; inorganic clay of high plasticity		
ASTM classification (See Figure 2.7)	CH		
Significant properties	≧50% passes No. 200 sieve (0.003 in.) $w_L > 50\%$ Atterberg plot above A line $I_p > 20$		
Field identification	Fine-grained soil of high plasticity; sticky and highly moldable without fracture when wet; non-draining; impervious; dry clumps very hard and very difficult to pulverize; highly compressible		
Average properties			
Allowable bearing (lb/ft^2)	Soft	Medium	Stiff
with minimum of one ft surcharge	500	750	1000
increase for surcharge (%/ft)	10	10	10
maximum total	1500	1500	1500
Lateral pressure			
active coefficient	0.75	0.85	0.95
passive (lb/ft^2 per ft depth)	33	100	167
Friction (coefficient or lb/ft^2)	200	300	400
Weight (lb/ft^3)			
dry	75	90	105
saturated	95	110	130
Compressibility	Very high	High	Medium high

TABLE 2.5. *(Continued)*

Organic silt and organic silty clay of low plasticity			Inorganic silt; micaceous or diatomaceous fine sands or silt; elastic silt		
OL			MH		
≧50% passes No. 200 sieve (0.003 in.) $w_L \leqq 50\%$ Atterberg plot below A line $I_p < 20$			≧50% passes No. 200 sieve (0.003 in.) $w_L > 50\%$ Atterberg plot below A line $I_p < 20$		
Fine-grained soil of low plasticity; slow draining; dry clumps quite hard, but not very difficult to pulverize; typical slight musty, rotting odor			Fine-grained soils of low plasticity; slow draining; dry clumps quite hard, but not very difficult to pulverize; spongy; compressible wet or dry		
Loose or soft ($N < 10$)	Medium ($10 < N < 30$)	Dense or stiff ($N > 30$)	Loose or soft ($N < 10$)	Medium ($10 < N < 30$)	Dense or stiff ($N > 30$)
500	750	1000	500	750	1000
10	10	10	10	10	10
2000	2000	2000	1500	1500	1500
0.75	0.85	0.95	0.50	0.60	0.75
33	50	67	33	50	67
250	375	500	200	300	400
75	90	100	70	85	100
95	105	115	100	110	120
High	Medium high	Medium	Very high	High	Medium high

TABLE 2.5. *(Continued)*

Organic clay of medium to high plasticity			Peat, muck, topsoil
OH			Pt
$\geqq$50% passes No. 200 sieve (0.003 in.) $w_L > 50\%$ Atterberg plot below A line			Highly organic Low density
Fine-grained soil of medium to high plasticity; sticky and moderately moldable without fracture when wet; non-draining; impervious; dry clumps hard; moderately difficult to pulverize; highly compressible; typical slight musty, rotting odor			Contains large amounts of partially decomposed plant or animal materials; strong rotting odor; slow draining highly compressible
Soft	Medium	Stiff	
500	500	500	Not usable
0	0	0	
500	500	500	
0.75	0.85	0.95	0.30
33	33	33	Not usable
150	200	200	Not usable
65	85	100	70–90
90	110	125	90–110
Very high	High	Medium high	Very high

Active lateral pressure. This is the horizontal pressure exerted against retaining structures, visualized in its simplest form as an equivalent fluid pressure. Quantification is in terms of a density for the equivalent fluid given in actual unit weight value or as a percentage of the soil unit weight.

Passive lateral pressure. This is the horizontal resistance offered by the soil to forces against the soil mass. It is also visualized as varying linearly with depth in the manner of a fluid pressure. Quantification is usually in terms of a specific pressure increase per unit of depth.

Friction resistance. This is the resistance to sliding along the contact bearing face of a footing. For cohesionless soils it is usually given as a friction coefficient to be multiplied by the compression force. For clays it is given as a specific value in pounds per square foot to be multiplied by the contact area.

Calculation of settlements is quite complex and is beyond the scope of this book. It is well to be generally aware of the conditions that may produce settlement problems, but the actual quantification of settlement values should be done by an experienced soils engineer.

Whenever possible, stress limits should be established as the result of a thorough investigation and the recommendations of a qualified soils engineer. Most building codes allow for the use of so-called *presumptive* values for design. These are average values, on the conservative side usually, that may be used for soils identified by groupings used by the codes. Reprints of portions of the *Uniform Building Code*, 1979 edition, and the *Building Code of the City of Los Angeles*, 1976 edition, are given in the appendix; both contain presumptive values for design. Soil types are identified only rather broadly in the *Uniform Building Code*, whereas the Los Angeles code uses what is essentially the unified system in establishing allowable bearing pressures.

Table 2.5 presents a summary of information for the basic soil types classified by the unified system. Information is grouped as follows:

Significant properties. These are the properties that relate directly to the identity or stress-and-strain evaluation of the soil.

Simple field identification. These are the various characteristics of the soil that may be used for identification in the absence of testing. They may also be used to verify that the soil encountered during construction is that indicated by the soil exploration and assumed for the foundation design.

Average properties. These are approximate values and are best verified by exploration and testing, but should be adequate for preliminary design. Stress values are generally those in the approximate range recommended by building codes.

Study Aids

Words and Terms. Using the glossary and the text of the preceding chapter for reference, review the meaning of the following words and terms.

Angle of internal friction (ϕ).
Atterberg limits.
Bedrock.
Clay.
Coarse fraction.
Cohesionless.
Cohesive.
Compaction.
Compressibility.
Consistency.
Consolidation.
Cut.
Density.
Erosion.
Excavation.
Expansive soil.
Fill.
Fines.
Flocculent soil structure.
Frost heave.
Frost line.
Gap-graded soil.
Grain size.
Grain shape.
Gravel.
Liquid limit (w_L).
Maximum density.
Penetration resistance (N).
Permeability.
Plastic limit (w_L).
Plasticity index (I_p).
Porosity (n).
Preconsolidation.
Presumptive bearing pressure.
Rock.
Sand.
Shrinkage limit (w_s).
Silt.
Soil.
Specific gravity (G).
Surcharge.
Unconfined compressive strength (q_u).
Unified system for soil classification.
Viscosity.
Void ratio (e).

Questions

1. What two materials ordinarily take up the void in a soil?
2. What condition makes it possible to estimate the void ratio of ordinary soils when only the dry unit weight is known?
3. What does it mean to say that a clay is *fat?*
4. Besides grain size, what single property is most important in distinguishing between silts and clays?
5. What condition qualifies a sand or gravel as *clean?*

Problems

See selected answers following the appendix.

1. Given the value for the dry unit weight, and assuming a specific gravity of 2.65, find the following for the samples of sand listed.
   a. Void ratio.
   b. Soil unit weight and water content if fully saturated.
   c. Water content if wet sample weighs 110 lb/ft^3.

Sample	Unit Dry Weight (lb/ft^3)
A	90
B	95
C	100
D	105

2. Given the value for the saturated unit weight, and assuming a specific gravity of 2.7, find the following for the samples of clay listed.
   a. Void ratio.
   b. Water content of the saturated sample.
   c. Dry unit weight of the sample.

Sample	Unit Saturated Weight (lb/ft^3)
A	90
B	100
C	110
D	120

3. Given the following data from the grain size analysis, classify the following soil samples according to general type (sand or gravel) and nature of gradation (uniform, well-graded, gap-graded).

Sample	D_{10} (mm)	D_{30} (mm)	D_{60} (mm)
A	1.00	4.00	50.00
B	0.20	1.00	2.00
C	0.10	0.25	0.40
D	0.06	0.15	7.00

4. Given the information listed for each soil sample, identify the soil using the unified system and estimate the values for allowable bearing pressure, lateral passive resistance, and friction resistance using Table 2.5.

   a. Grain size analysis: 60% retained on No. 4 sieve, only 4% passes No. 200 sieve, $D_{10} = 0.4, D_{30} = 3, D_{60} = 8$.
   Penetration resistance: $N = 20$.

   b. Grain size analysis: 80% passes No. 4 sieve, 6% passes No. 200 sieve, $D_{10} = 0.1, D_{30} = 0.4, D_{60} = 2$.
   Penetration resistance: $N = 8$.

   c. In this case, 60% passes No. 200 sieve, liquid limit = 40%, plasticity index = 10, unconfined compressive strength: $q_u = 1.6$ k/ft^2. Strong musty odor.

3

Design of Shallow Bearing Foundations

Shallow foundation is the term usually used to describe the type of foundation that transfers vertical loads by direct bearing on soil strata close to the bottom of the building and a relatively short distance below the ground surface. There are three basic forms of shallow foundations: the continuous, strip wall footing; the individual column or pier footing; and the mat or raft foundation. A raft is produced by turning the entire bottom of the building into one large footing. The design of raft foundations is beyond the scope of this book. The behavior of bearing footings relating to soil properties and soil stress mechanisms is discussed in Chapter 2. The material in this chapter deals with the design of wall and column footings and various elements of foundation construction such as foundation walls, grade beams, ties and piers that are ordinarily used in conjunction with shallow bearing systems.

Figure 3.1 illustrates a variety of elements ordinarily used in bearing foundation systems. As noted in the illustration, the design of these elements is discussed in various portions of this book.

3.1 Foundation Walls

Foundation walls are walls that extend below the ground surface and perform some sort of transition between the foundation and the por-

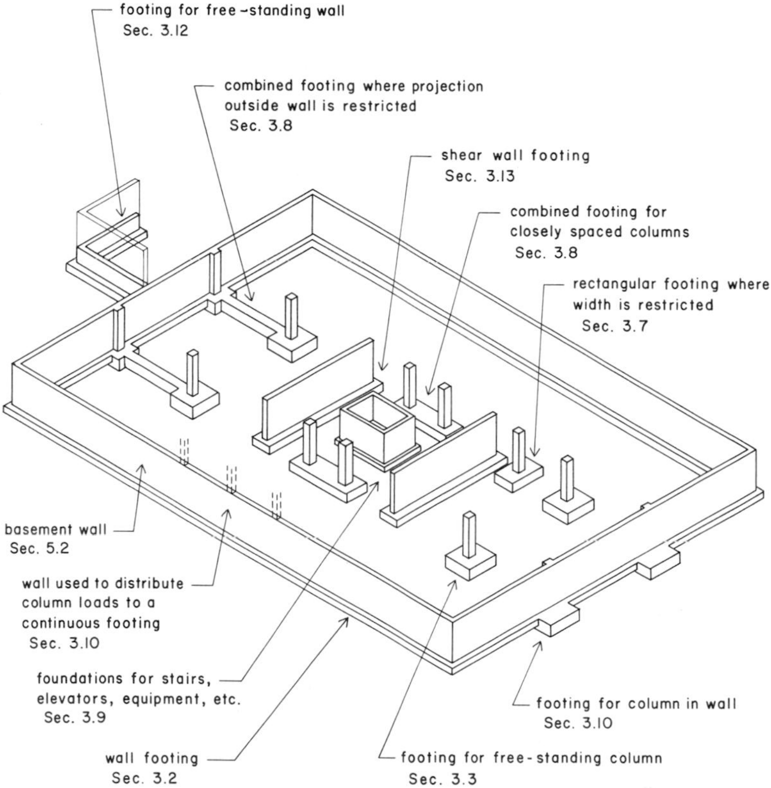

FIGURE 3.1. Ordinary elements of shallow bearing foundation systems.

tions of the building that are aboveground. They are typically built of concrete or masonry. The structural and architectural functions of foundation walls vary considerably, depending on the type of foundation, the size of the building, climate and soil conditions, and whether or not they serve to form a basement.

Figure 3.2 shows some typical situations for the use of foundation walls for buildings without basements. In these cases the walls are not actually walls in the usual architectural sense. A principal difference relates to the construction of the building floor, whether it consists of a framed structure elevated above the ground, or concrete placed directly on the ground. Another major difference has to do with the

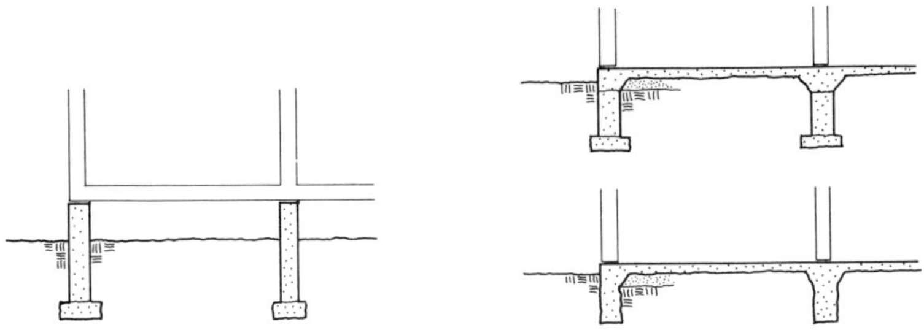

framed floor over a crawl space   concrete floor poured on the ground

FIGURE 3.2. Typical foundations for bearing walls—buildings without basements.

distance of the foundation below the ground surface. If this distance is great, because of the need for frost protection or the need to reach adequate soil for bearing, the walls may be quite high. If these problems do not exist, the walls may be quite short, and scarcely constitute walls at all in the usual sense, especially when there is no crawl space, as shown in Figure 3.2.

When the wall is very short and building loads are low, it is sometimes possible to use a construction known as a grade beam. This consists of combining the functions of foundation wall and wall footing into a single element that provides continuous support for elements of the building construction.

When a basement is required foundation walls are usually quite high. An exception is the case of a half-basement in which the basement floor is only a short distance below grade and the portion of the basement wall aboveground may be of different construction than that extending into the ground. If the basement floor is a significant distance below grade, major soil pressure will be exerted horizontally against the outside of the wall. In such a case the wall will function as a spanning element supported laterally by the footing or the basement floor at its lower end and by the building floor at its top. The design of basement walls is discussed in Section 5.2.

Figure 3.3 shows a number of situations in which foundation walls function to provide basement spaces. For buildings with multilevel basements, walls may become quite massive due to the accumulation of vertical load and the potential for considerable lateral soil pressure.

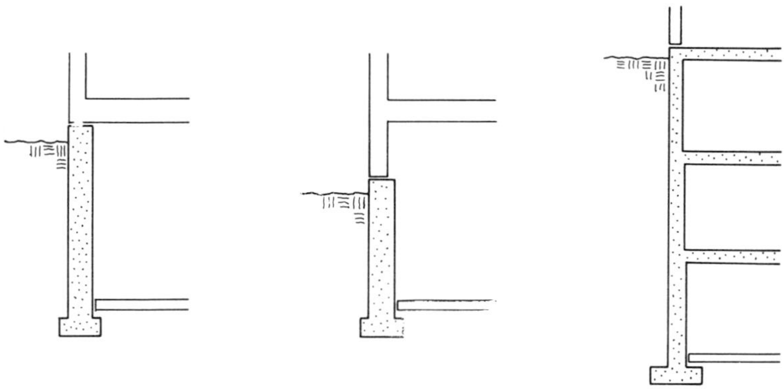

FIGURE 3.3. Foundation walls for buildings with basements.

In addition to their usual functions of providing a ground-level edge for the building and support for elements of the building, walls often serve a variety of functions for the building foundation system. Some of these are as follows:

Load distributing or equalizing. Walls of some length and height typically constitute rather stiff beamlike elements. Their structural potential in these cases is often utilized for load distributing or equalizing, as shown in Figure 3.4. Even the shallow grade beam is typically designed with continuous top and bottom reinforcing to serve as a continuous beam for these purposes; hence the derivation of its name.

Spanning as a load-carrying beam. Walls may be used as spanning members, carrying their own weight as well as some supported loads, as shown in Figure 3.5. This is often the case in buildings with deep foundations and a column structure. Deep foundation elements are placed under the columns, and walls are used to span from column to column. This can also be the situation with bearing foundations if the column footings are quite large and reasonably closely spaced; rather than bearing on its own small footing, the stiff wall tends to span between the larger footings.

Distribution of column loads. When columns occur in the same plane as a foundation wall, many different relationships for the structural

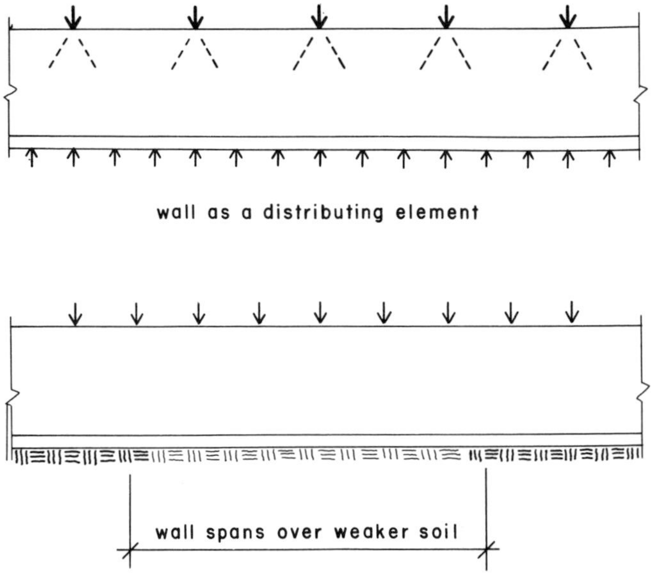

FIGURE 3.4. Load distributing and spanning effects of continuous foundation walls.

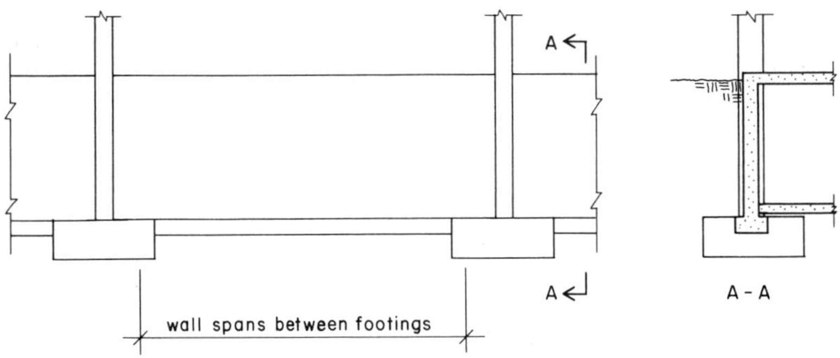

FIGURE 3.5. Spanning action of continuous foundation wall in line with large columns.

action of the walls, columns, and column foundations are possible. If the walls and columns are monolithically constructed, some load sharing is generally unavoidable. The various possibilities are discussed in Section 3.10.

Transfer of building lateral loads to the ground. Although there are many possible situations for the development of lateral resistive structural systems, the total lateral load on the building must ultimately be transferred to the ground. The horizontal force component is usually transferred through some combination of soil friction on the bottoms of footings and the development of passive lateral soil pressure against the sides of the footings and foundation walls. For large buildings with shallow below-grade structures this may be a major task for the foundations walls.

Ties, struts, collectors, etc. Foundation walls may be used to push or pull forces between separate elements of the below-grade structure. For seismic design it is usually required that the separate elements of the foundation be adequately tied to permit them to move as a single mass; where they exist, foundation walls may help to serve this purpose.

If a foundation wall is sufficiently tall, it should be treated as a wall and reinforced in both directions with minimal reinforcing for shrinkage and temperature stresses, as recommended by the ACI code (Ref. 10).

3.2 Wall Footings

Wall footings consist of concrete strips placed under walls. The most common type of wall footing is that shown in Figure 3.6, consisting of a strip with a rectangular cross section placed in a symmetrical position with respect to the wall and projecting an equal distance as a cantilever from both faces of the wall. For soil stress the critical dimension of the footing is the width of the footing bottom measured perpendicular to the wall face.

In most situations the wall footing is utilized as a platform upon which the wall is constructed. Thus a minimum width for the footing is established by the wall thickness, the footing usually being made somewhat wider than the wall. With a concrete wall this additional width is

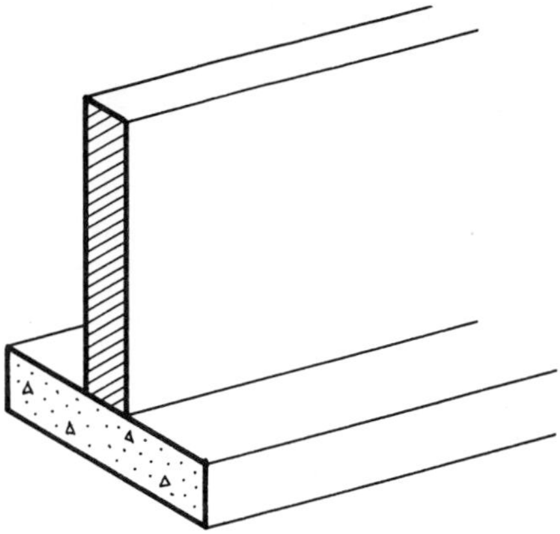

FIGURE 3.6. Typical freestanding wall and its footing.

used to support the wall forms while the concrete is poured. For masonry walls this added width assures an adequate base for the mortar bed for the first course of the masonry units. The exact additional width required for these purposes is a matter of judgment. For support of concrete forms it is usually desirable to have at least a 3 in. projection; for masonry the usual minimum is 2 in.

With relatively lightly loaded walls the minimum width required for platform considerations may be more than adequate in terms of the allowable bearing stress on the soil. If this is the case, the short projection of the footing from the wall face will produce relatively insignificant transverse bending and shear stresses, permitting a minimal thickness for the footing and the omission of transverse reinforcing. Most designers prefer, however, to provide some continuous reinforcing in the long direction of the footing, even when none is used in the transverse direction. The purpose is to reduce shrinkage cracking and also to give some enhanced beamlike capabilities for spanning over soft spots in the supporting soil.

As the wall load increases the increased width of the footing required to control soil stress eventually produces significant transverse bending and shear in the footing. At some point this determines the required thickness for the footing and for required reinforcing in the transverse direction. If the footing is not reinforced in the transverse direction, the controlling stress is usually the transverse tensile bending stress in the concrete. If the footing has transverse reinforcing, the controlling concrete stress is usually the shear stress.

A number of different situations for wall footings are shown in Figures 3.7 through 3.10. In mild climates where frost protection is not critical and where soil conditions permit it, footings are often placed a very short distance below the ground surface. Figure 3.7 shows details of typical wall footings used in this situation. When the building has no basement and the floor consists of a concrete paving slab placed directly on the ground, these footings are sometimes poured monolithically with the floor slab.

Paving slabs are discussed in detail in Section 6.1. The details shown in Figure 3.7 assume a thin concrete slab poured on a compacted layer of granular fill. This type of slab is typically minimal in thickness and slightly reinforced, usually with a single thickness of welded wire fabric. Since the footings tend to provide a stiffer support than the adjacent soil, some designers prefer to provide some reinforcing bars in the top of the slab at the footing to minimize tension cracking.

The heavy lines in the sections in Figure 3.7 indicate the profile of the excavation. For very shallow footings the excavation is sometimes accomplished by first performing a general excavation to the level of the bottom of the granular fill and then trenching additionally as required to form the bottoms of the footings, as shown in Figure 3.8. If this is done, constructed forming for the footings is limited to that

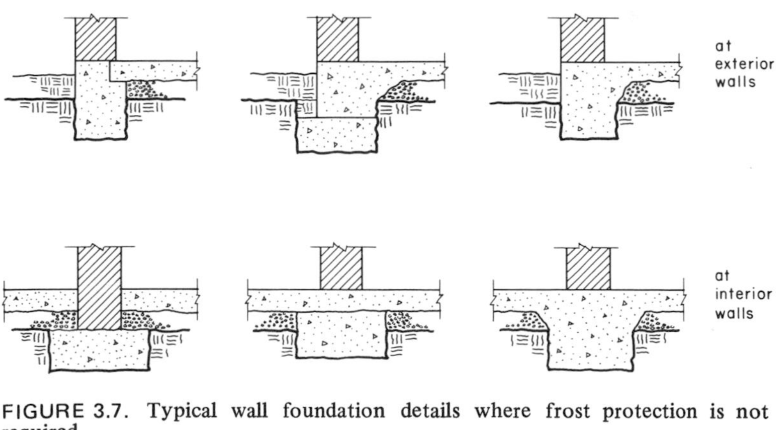

FIGURE 3.7. Typical wall foundation details where frost protection is not required.

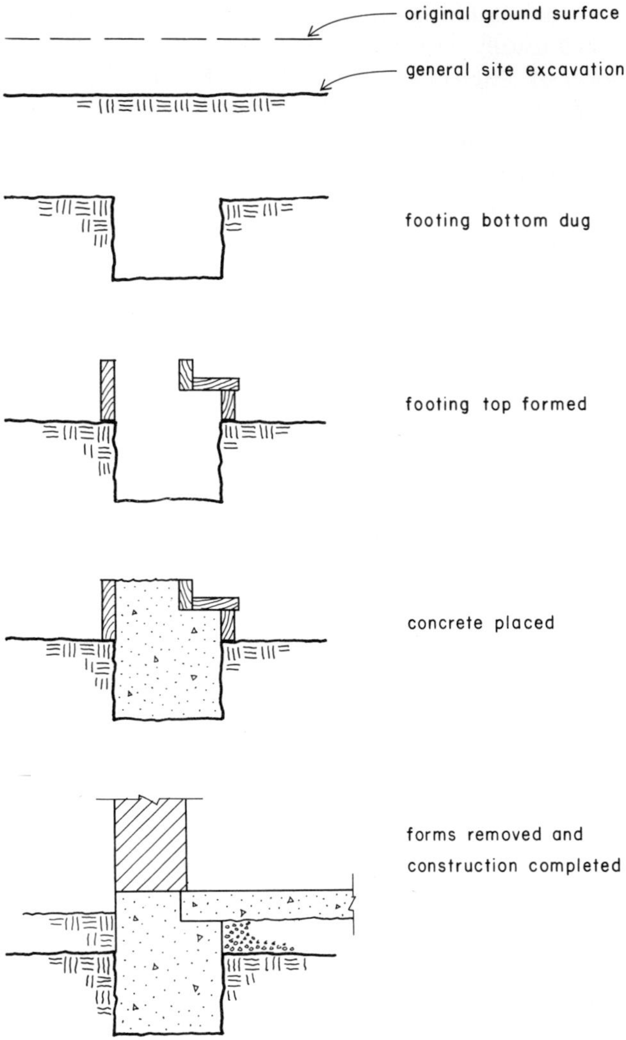

FIGURE 3.8. Typical construction sequence for a simple shallow wall footing.

required above the level of the general excavation. This practice is most feasible with reasonably cohesive soils that can maintain a vertical cut on the sides of the excavation.

When footings must be placed a greater distance below the ground surface because of frost conditions or the presence of weak upper soil strata, the foundation usually consists of a footing and a separately poured foundation wall. Examples of this type of construction are shown in Figure 3.9. At (*a*) is shown a modification of the shallow footing shown in Figure 3.7. The detail in (*b*) would be used for deeper footings with the foundation wall poured before the fill is placed for the floor slab. Detail (*c*) shows the typical construction used with a crawl space; it is essentially the same as in (*b*) so far as the wall and footing are concerned.

Figure 3.10 shows a typical exterior basement wall with a full basement and wood frame construction. In addition to the usual functions of bearing and serving as a construction platform, the footing may need to provide lateral support for the bottom of the wall if the backfill on the outside of the wall is placed before pouring the basement floor slab. If this is the case, a keyway in the top of the footing is required to keep the wall from slipping off the footing, although this detail is often used as a matter of habit.

Another type of wall footing is that provided for a wall that is freestanding, that is, not laterally supported at its top. If this type of wall is outdoors, it must sustain horizontal wind pressure as well as gravity loads. In high-risk seismic zones such a wall must sustain horizontal

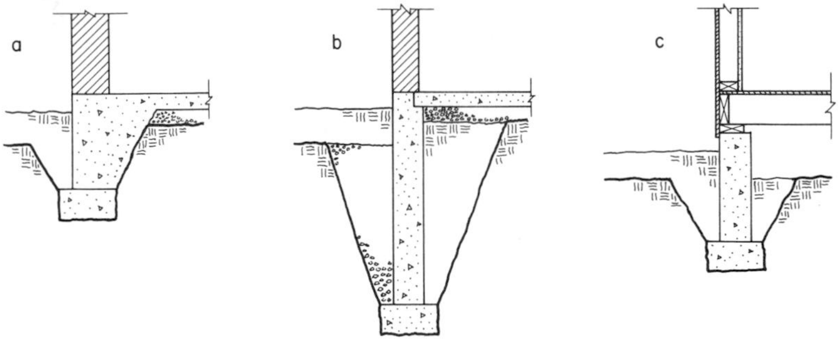

FIGURE 3.9. Wall foundations used where a deeper footing is required for frost protection.

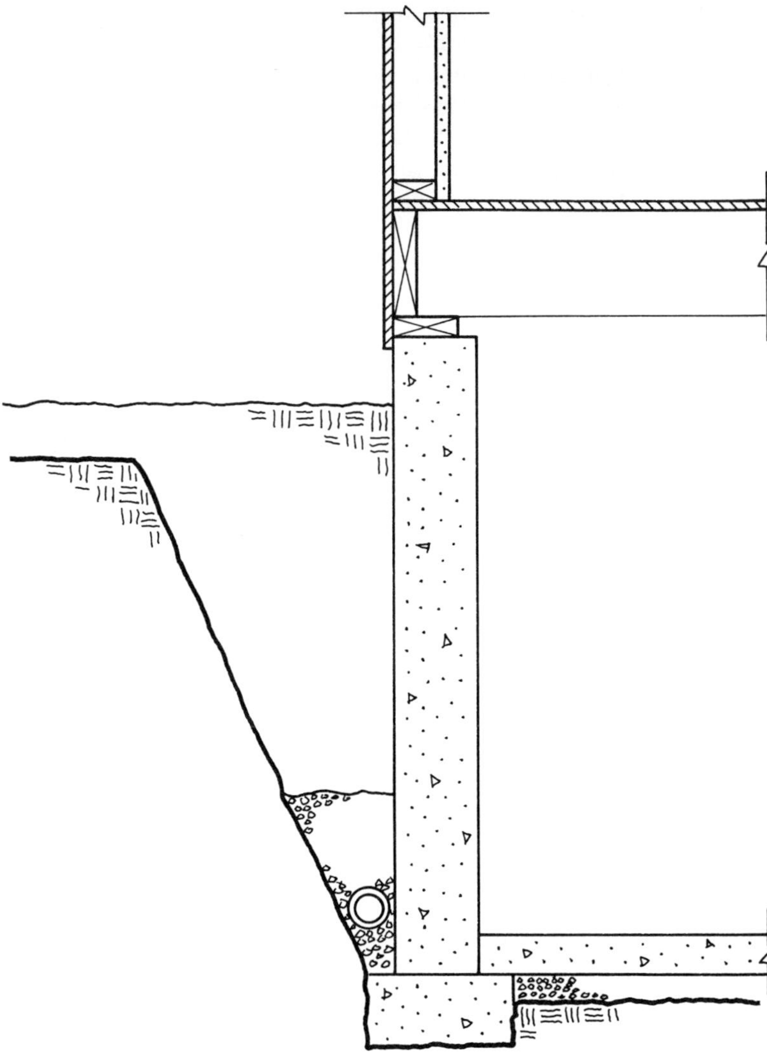

FIGURE 3.10. Typical basement foundation wall.

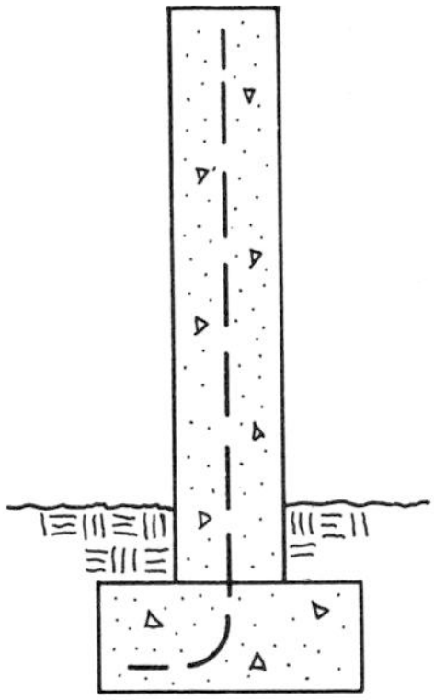

FIGURE 3.11. Typical symmetrical footing for a freestanding wall.

seismic load whether outdoors or indoors. If there is a difference in grade levels on the two sides of such a wall, it must also be designed for the lateral load of the earth pressure on the high side.

Figure 3.11 shows a simple freestanding wall such as may be used for a garden wall or property line fence. The footing for such a wall is typically a symmetrically placed element as shown. The footing width required beyond the usual platform requirement is generally determined by the lateral loads, if the wall is of significant height. Foundations for freestanding walls are discussed in Section 3.12. The general problems and design of retaining walls are discussed in Chapter 5.

Details for wall footings are often determined primarily by the general requirements of the building construction and some simple, pragmatic considerations of the excavation and construction work. Some building codes provide minimum requirements for light construction, such as that shown in Figure 3.12. In these cases the foundation design may involve little more than the determination of the loads on the footings to verify that the required minimum widths are adequate for the maximum allowable soil stress. When loads are significantly higher than those that can be sustained by the minimal footings, some structural calculations are required to assure a safe limit on concrete stresses and to determine any required reinforcing. The material that follows illustrates the design procedures for typical wall footings. Calculations for reinforced concrete are done by the working stress method. For a digest of the procedures and criteria used for this method see the appendix.

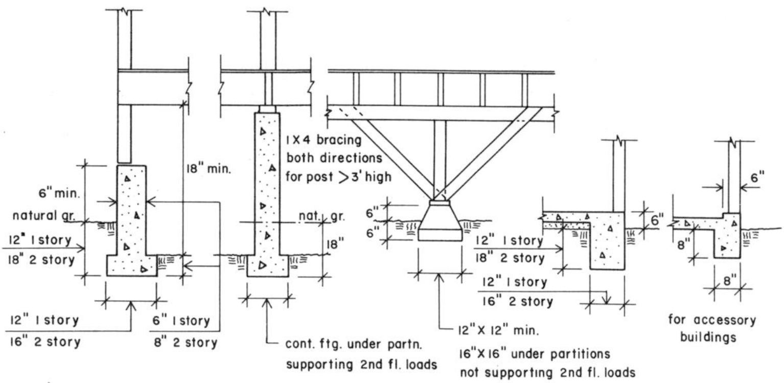

FIGURE 3.12. Typical recommended details for foundations for light wood frame construction. Adapted from the information sheet for Type V construction issued by the Department of Building and Safety of the City of Los Angeles.

The design of a symmetrical wall footing consists of the following:

1. ***Determination of the footing width.*** Footing width is determined by the soil pressure limit, assuming that the minimum width required for construction is not adequate for bearing. Since the weight of the footing is part of the total load on the soil, the required width cannot be precisely determined until the footing thickness is known. The usual design procedure is to estimate a required footing thickness, find the corresponding width required, check the footing stresses to verify the adequacy of the thickness, and–if necessary–revise the thickness and find a new required width. In the examples in this section we will use the procedure of designing for the allowable maximum soil pressure under total design load on the footing. The procedure for designing for equalization of dead load pressures in a series of footings is illustrated in Section 3.15.

2. ***Determination of the footing thickness.*** If the footing has no transverse reinforcing, the thickness is determined by the tension stress limit of the concrete, in either flexural bending stress or diagonal stress due to shear. If the footing has transverse reinforcing, the stress limits become the flexural compressive stress and

diagonal tension stress in the concrete and the flexural tension stress and bond stress on the reinforcing. Since the reinforcing is quite costly, whereas concrete poured into a hole in the ground is relatively inexpensive, the footing thickness is usually not critical with regard to compressive strength of the concrete. Instead, a thickness is chosen that is adequate to eliminte the need for shear reinforcing and to minimize the percentage of tensile reinforcing. Minimum footing thicknesses are a matter of judgment, unless limited by building codes. The ACI code (Ref. 11) recommends a minimum of 8 in. for an unreinforced footing and 10 in. for one with transverse reinforcing. Another possible consideration for the minimum footing is the necessity for placing dowels for the wall reinforcing.

3. ***Selection of reinforcing.*** Transverse reinforcing is selected on the basis of the tension and bond stresses generated in the cantilever action of the footing. Longitudinal reinforcing is usually selected on the basis of providing minimal shrinkage reinforcing. Although no specific criteria are established for the latter situation, a reasonable minimum value is a steel area of 0.0015 times the gross concrete area, which is the minimum reinforcement required in a horizontal direction in walls by the ACI code (Ref. 11). Cover, or the distance of the reinforcing from the edges of the footing, must be a minimum of 3 in. if the edge is unformed and 2 in. if the edge is formed. Except for very heavily stressed foundations, reinforcing bars are usually of a low grade of steel, often the lowest grade available as deformed reinforcing bars.
4. ***Development of details.*** Some of the details required are the placing of reinforcing, construction joints between the footing and walls and slabs, forming of the aboveground portions of footings, and such possible features as keyways, waterstops, and openings or sleeves for pipes or conduits.

The following examples illustrate procedures for the design of unreinforced and reinforced wall footings. Tabulations of predesigned footings are given in Tables 3.1 and 3.2. The table entries were determined by the procedures illustrated in the examples. We do not recommend the use of unreinforced footings more than 3 ft wide.

TABLE 3.1. Allowable Loads on Unreinforced Wall Footings[a]

[a] For illustration and discussion of the process used to generate the table data see Example 1 in the text. Allowable loads do not include the weight of the footing, which has been deducted from the total bearing capacity. Example: 1000 lb/ft² pressure, 6 in. thick by 18 in. wide footing; total capacity = 1000(1.5) = 1500 lb, footing weighs 75 lb/ft², allowable load = (1000 – 75) (1.5) = 1388 lb/ft.

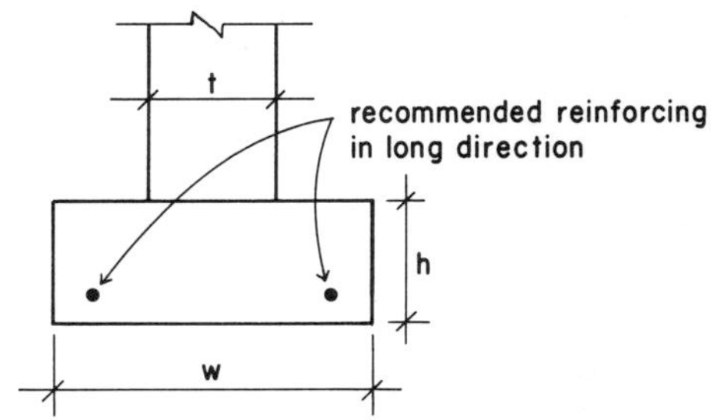

Maximum Soil Pressure (lb/ft²)	Minimum Wall Thickness: Concrete t (in.)	Masonry t (in.)	$f'_c = 2000$ psi: Allowable Load on Footing (lb/ft)	h (in.)	w (in.)	Reinforcing Options	$f'_c = 3000$ psi: Allowable Load on Footing (lb/ft)	h (in.)	w (in.)	Reinforcing Options
1000	4	8	925	6	12	1 No. 3	(Note) absence of entry under this section denotes that data is same as for $f'_c = 2000$ lb/in.²			
	4	8	1156	6	15	2 No. 3				
	4	8	1388	6	18	2 No. 3				
	6	12	1850	6	24	2 No. 3				
	6	12	2280	7	30	3 No. 3, 2 No. 4	2312	6	30	3 No. 3, 2 No. 4
	6	12	2700	8	36	3 No. 4, 2 No. 5	2736	7	36	3 No. 4, 2 No. 5

1500	4	8	1425	6	12	1 No. 3				
	4	8	1781	6	15	2 No. 3				
	4	8	2138	6	18	2 No. 3				
	6	12	2850	6	24	2 No. 3				
	6	12	3500	8	30	2 No. 4	3531	7	30	3 No. 3, 2 No. 4
	6	12	4125	10	36	3 No. 4, 2 No. 5	4162	9	36	3 No. 4, 2 No. 5
2000	4	8	1925	6	12	1 No. 3				
	4	8	2406	6	15	2 No. 3				
	4	8	2888	6	18	2 No. 3				
	6	12	3824	7	24	3 No. 3, 2 No. 4				
	6	12	4719	9	30	4 No. 3, 2 No. 4				
	6	12	5588	11	36	3 No. 4, 2 No. 5	5625	10	36	3 No. 4, 2 No. 5
3000	4	8	2925	6	12	1 No. 3				
	4	8	3656	6	15	2 No. 3				
	4	8	4369	7	18	2 No. 3	4388	6	18	2 No. 3
	6	12	5775	9	24	3 No. 3, 2 No. 4	5800	8	24	3 No. 3, 2 No. 4
	6	12	7156	11	30	3 No. 4, 2 No. 5	7188	10	30	3 No. 4, 2 No. 5
	6	12	8475	14	36	4 No. 4, 3 No. 5	8512	13	36	4 No. 4, 3 No. 5

EXAMPLE 1 UNREINFORCED WALL FOOTING

Figure 3.13 shows a simple symmetrical wall footing of indefinite length. Design of this type of element is typically done on a "strip" basis, as for a concrete slab or wall. The criteria for the design is as follows:

Footing design load: 4800 lb/ft of wall length.

Wall thickness for design: 6 in.

(This is the actual thickness for a concrete wall or half the actual thickness for a masonry wall; see Section 2304 of the ACI code–Ref. 10.)

Maximum allowable soil pressure: 2000 lb/ft^2.

Concrete design strength: $f'_c = 2000$ psi.

The two concrete stress conditions affecting the design are the bending and shear produced by the cantilevering action of the footing from the

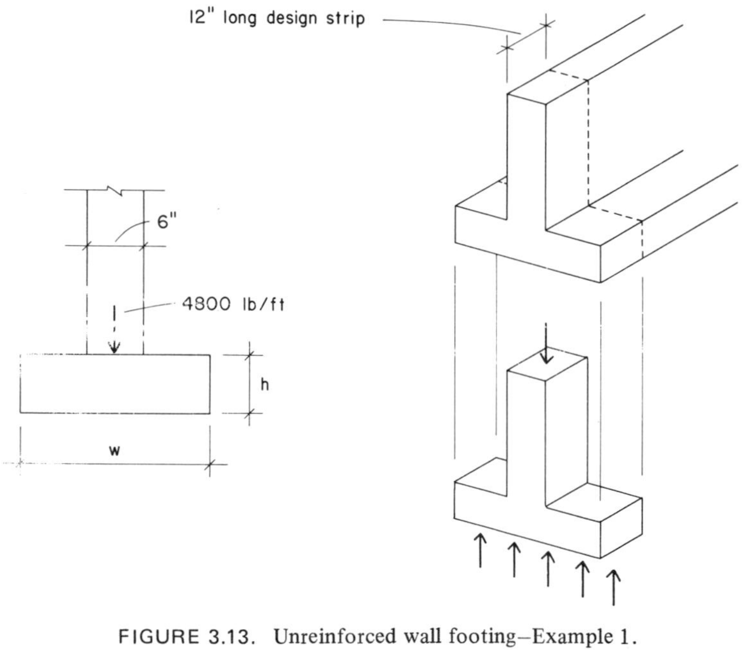

FIGURE 3.13. Unreinforced wall footing–Example 1.

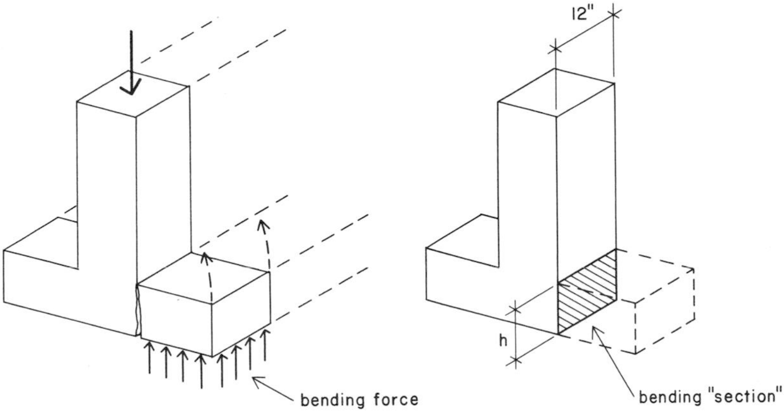

FIGURE 3.14. Bending action in the unreinforced wall footing–Example 1.

face of the wall. The bending action and the critical section for bending resistance are shown in Figure 3.14. The allowable stress for this situation is that permitted for flexure in tension with plain (unreinforced) concrete, as given in Table 1002 (*a*) of the ACI code: $f_c = 1.6\sqrt{f_c'}$.

For shear in this case the actual critical condition is that of the resultant diagonal tension stress. Thus the probable mode of failure is of a form usually assumed for so-called punching shear, as illustrated in Figure 3.15. This is the basis for analysis of shear in concrete in

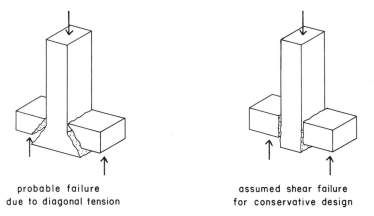

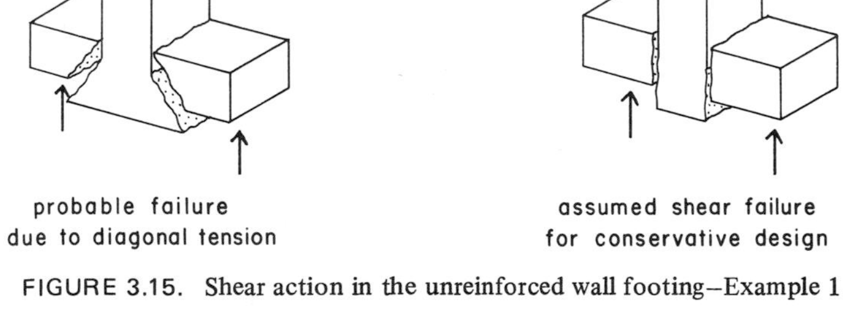

FIGURE 3.15. Shear action in the unreinforced wall footing–Example 1.

reinforced footings. However, we recommend a slightly more conservative assumption of failure by assuming a vertical shear failure section at the face of the wall; the same section used for flexure. For the allowable shear we use that given for shear as a measure of diagonal tension on the web of a beam without shear reinforcing: $v_c = 1.1\sqrt{f'_c}$. In any event, the shear assumption is usually academic, since the flexural stress is ordinarily the limiting condition for design.

As has been discussed previously, the footing width must be selected on the basis of the total soil pressure, part of which is due to the weight of the footing itself and therefore cannot be determined until the footing thickness is known. The usual procedure is to guess at a thickness, determine the required width, and then check the footing for stress. If the first guess is very far off, a second try is made. This process is easier, of course, if one has access to some previous footing designs using the same soil and concrete stress criteria, such as those tabulated in Table 3.1.

Try: $h = 10$ in.

Then: footing weight $= \frac{10}{12}(150) = 125$ lb/ft^2

usable soil pressure $= 2000 - 125 = 1875$ lb/ft^2

required $w = \dfrac{4800}{1875} = 2.56$ ft or 30.72 in.

Try: $w = 31$ in.

Then: actual soil pressure $= \dfrac{4800 \times 12}{31} = 1858$ lb/ft^2 (for concrete stress)

With the 31 in. wide footing the cantilever distance from the face of the wall is 12.5 in. and the total soil pressure force on the cantilever is thus

$$F = (1858)\left(\frac{12.5}{12}\right) = 1935 \text{ lb}$$

and the cantilever moment, as shown in Figure 3.16, is

$$M = (1935)\left(\frac{12.5}{2}\right) = 12{,}094 \text{ lb-in.}$$

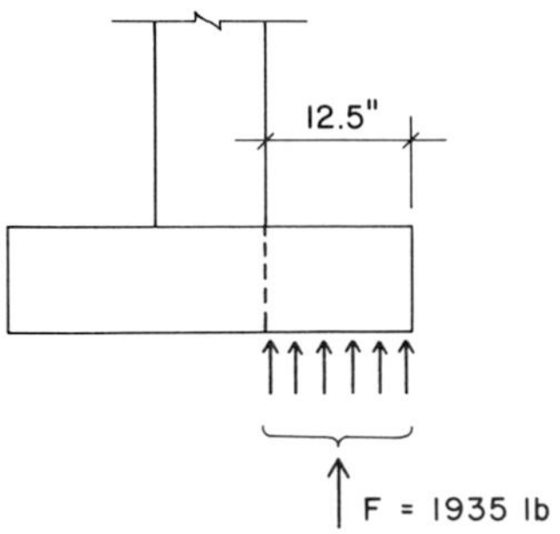

FIGURE 3.16. Cantilever action of the wall footing–Example 1.

The section that resists this stress is the 12 in. wide by 10 in. high section for which the section modulus is

$$S = \frac{(12)(10)^2}{6} = 200 \text{ in.}^3$$

and the maximum bending stress is thus

$$f = \frac{M}{S} = \frac{12{,}094}{200} = 61 \text{ psi}$$

which is compared to the allowable stress of

$$f_c = 1.6\sqrt{f'_c} = 1.6\sqrt{2000} = 72 \text{ psi}$$

This indicates that the first try for the thickness is quite conservative. If we reduce the thickness to 9 in., a second try would proceed as follows:

$$\text{new footing weight} = \tfrac{9}{12}(150) = 112.5, \text{ say } 113 \text{ lb/ft}^2$$

$$\text{usable soil pressure} = 2000 - 113 = 1887 \text{ lb/ft}^2$$

$$\text{required } w = \frac{4800}{1887} = 2.54 \text{ ft or } 30.52 \text{ in.}$$

Thus the footing width does not change and the soil pressure, bending force, and bending moments will be the same as those determined previously. The section modulus, however, reduces to

$$S = \frac{(12)(9)^2}{6} = 162 \text{ in.}^3$$

and the new maximum bending stress becomes

$$f_c = \frac{12{,}094}{162} = 75 \text{ psi}$$

which exceeds the allowable stress of 72 psi.

Thus the 10 in. thick footing is the thinnest allowable, if whole-inch units of change are used. Actually, if the edges of the footing are board-formed, rather than trenched, logical thickness dimensions are those of actual lumber sizes. In this event, the best choice would be 9.25 in., which is the actual width of a nominal 2-by-10 and will result in a bending stress of 71 psi.

With our conservative assumption for shear stress, the shear force on the section will be the same as the bending force: 1935 lb as previously calculated. Thus the shear stress (v_c) is determined as follows on the 10 in. thick footing.

$$v_c = \frac{1935}{(12)(10)} = 16 \text{ psi}$$

which is well below the allowable stress of

$$v_c = 1.1\sqrt{f'_c} = 1.1\sqrt{2000} = 49 \text{ psi}$$

As discussed previously, even though no transverse steel reinforcing is used, we recommend a minimum longitudinal reinforcing for the wall footing consisting of a steel area of 0.0015 times the gross concrete cross section. For our 10 in. thick and 31 in. wide footing this steel area is

$$A_s = (0.0015)(10)(31) = 0.465 \text{ in.}^2$$

which may be satisfied by using three No. 4 bars with

$$A_s = (3)(0.20) = 0.60 \text{ in.}^2$$

The calculations in the preceding example illustrate the basis for determination of the footing designs given in Table 3.1. Note that the use of higher-strength concrete is not effective until the footing achieves significant width. The footing thicknesses given in the table are minimum ones determined on the basis of stresses assuming the full effectiveness of the footing cross section. If the footing concrete is poured on reasonably dry, porous soil, the concrete in contact with the soil will not be of a high quality because of the rapid absorption of water and cement by the adjacent soil. In this case it is recommended that the footing thickness be increased by 2 in. over that given in the table and the allowable loads be reduced accordingly. Designers should also be aware of the minimum footing thicknesses required by local building codes.

Entries in Table 3.1 are limited to footings 3 ft wide and maximum soil pressure of 3000 lb/ft^2. For wider footings or higher soil pressures we recommend the use of transverse reinforcing. The following example illustrates the design of such elements and is the basis for the determination of the entries in Table 3.2.

Example 2 Reinforced Wall Footing

Figure 3.17 shows the loading condition for this example. The design that follows is in general accord with the requirements of the latest ACI code (Ref. 10) but the calculations are done by the working stress criteria given in the 1963 edition of the code, portions of which are reprinted in the appendix. Criteria for the example are as follows:

Footing design load: 8750 lb/ft of wall length.
Wall thickness for design: 6 in.
Maximum soil pressure: 2000 lb/ft^2.
Concrete design strength: $f_c' = 2000$ psi.
Allowable tension on reinforcing: 20,000 psi.

For the reinforced footing the only concrete stress of concern is that in shear. Although concrete stress in compression is a potential concern, the footing thickness will usually be established by considerations of shear stress and the desire to minimize the amount of transverse reinforcing.

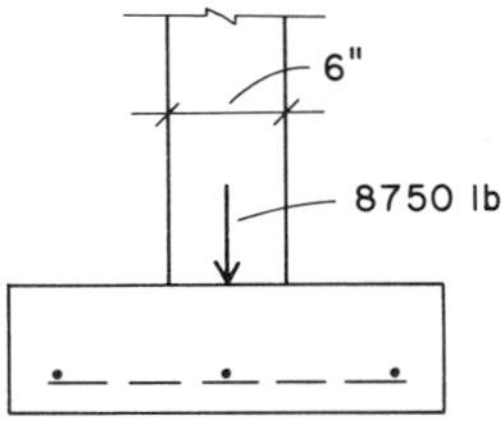

FIGURE 3.17. Reinforced wall footing–Example 2.

TABLE 3.2. Allowable Loads on Reinforced Wall Footings[a]

[a] $f_s = 20$ ksi, $v_c = 1.1\sqrt{f'_c}$, $f_c = 0.45\,f'_c$
Minimum concrete strength: $f'_c = 2000$ psi.
For illustration and discussion of the process used to generate the table data see Example 2 in the text. Allowable loads do not include the weight of the footing, which has been deducted from the total bearing capacity. Example: 10 in. thick by 48 in. wide footing, 1000 lb/ft^2 soil pressure; total capacity = (1000) (4) = 4000 lb / ft; footing weighs 125 lb/ft^2 ; allowable load = (1000 − 125) (4) = 3500 lb/ft.

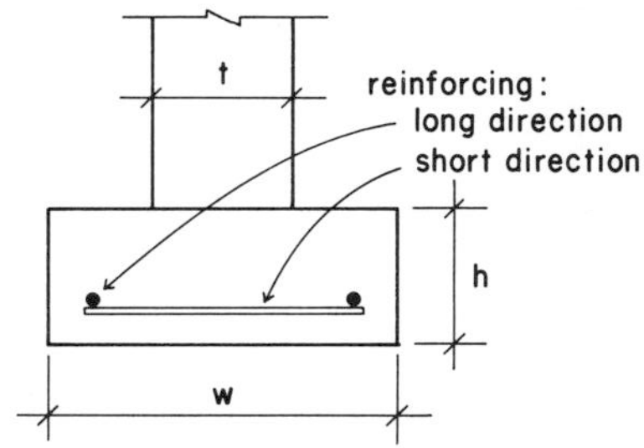

Maximum Soil Pressure (lb/ft^2)	Minimum Wall Thickness: Concrete t (in.)	Masonry t (in.)	Allowable Load on Footing (lb/ft)	Footing Dimensions: h (in.)	w (in.)	Reinforcing: Long Direction	Short Direction
1000	4	8	2625	10	36	3 No. 4	No. 3 at 16
	4	8	3062	10	42	2 No. 5	No. 3 at 12
	6	12	3500	10	48	4 No. 4	No. 4 at 16
	6	12	3938	10	54	3 No. 5	No. 4 at 13
	6	12	4375	10	60	3 No. 5	No. 4 at 10
	6	12	4812	10	66	5 No. 4	No. 5 at 13
	6	12	5250	10	72	4 No. 5	No. 5 at 11

1500	4	8	4125	10	36	3 No. 4	No. 3 at 10
	4	8	4812	10	42	2 No. 5	No. 4 at 13
	6	12	5500	10	48	4 No. 4	No. 4 at 11
	6	12	6131	11	54	3 No. 5	No. 5 at 15
	6	12	6812	11	60	5 No. 4	No. 5 at 12
	6	12	7425	12	66	4 No. 5	No. 5 at 11
	8	16	8100	12	72	5 No. 5	No. 5 at 10
2000	4	8	5625	10	36	3 No. 4	No. 4 at 14
	6	12	6562	10	42	2 No. 5	No. 4 at 11
	6	12	7500	10	48	4 No. 4	No. 5 at 12
	6	12	8381	11	54	3 No. 5	No. 5 at 11
	6	12	9250	12	60	4 No. 5	No. 5 at 10
	8	16	10106	13	66	4 No. 5	No. 5 at 9
	8	16	10875	15	72	6 No. 5	No. 5 at 9
3000	6	12	8625	10	36	3 No. 4	No. 4 at 10
	6	12	10019	11	42	4 No. 4	No. 5 at 13
	6	12	11400	12	48	3 No. 5	No. 5 at 10
	6	12	12712	14	54	6 No. 4	No. 5 at 10
	8	16	14062	15	60	5 No. 5	No. 5 at 9
	8	16	15400	16	66	5 No. 5	No. 6 at 12
	8	16	16725	17	72	6 No. 5	No. 6 at 10

As with the unreinforced footing, the usual design procedure consists of making a guess for a footing thickness and determining conditions to verify the guess.

Try: $h = 12$ in.

Then: footing weight = 150 lb/ft²

usable soil pressure = 2000 − 150 = 1850 lb/ft²

$$\text{required } w = \frac{8750}{1850} = 4.73 \text{ ft or } 56.8 \text{ in.}$$

Try: $w = 57$ in. or 4 ft − 9 in.

$$\text{Then: design soil pressure} = \frac{8750}{4.75} = 1842 \text{ lb/ft}^2$$

For the reinforced footing it is necessary to determine the effective depth of the cross section; that is, the distance from the top of the footing (compression face) to the center of the steel reinforcing. For precise calculation this involves a second guess: the steel bar diameter. As shown in Figure 3.18, we assume a No. 6 bar with $\frac{3}{4}$ in. diameter, which with the required 3 in. cover produces a net dimension for d as follows:

$$d = h - 3 - \frac{D}{2} = 12 - 3 - 0.375 = 8.625 \text{ in.}$$

Concern for precision is quite academic in footing design, however, considering the crude nature of the construction, so we will round off the value for d to 8.6 in. for the calculations. The code requires that the critical section for shear stress be taken at a distance d from the face of

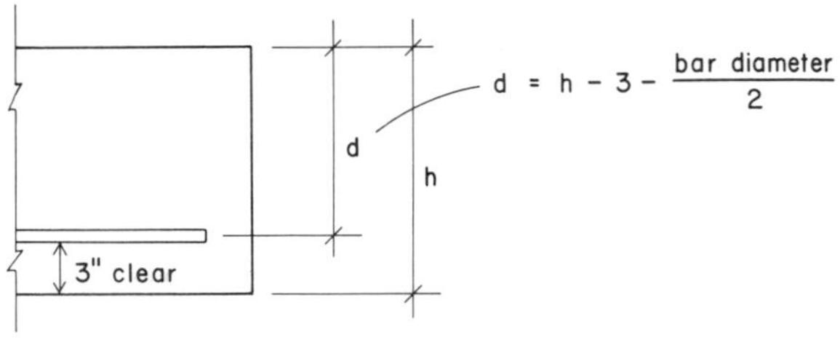

FIGURE 3.18. Effective depth for the reinforced wall footing.

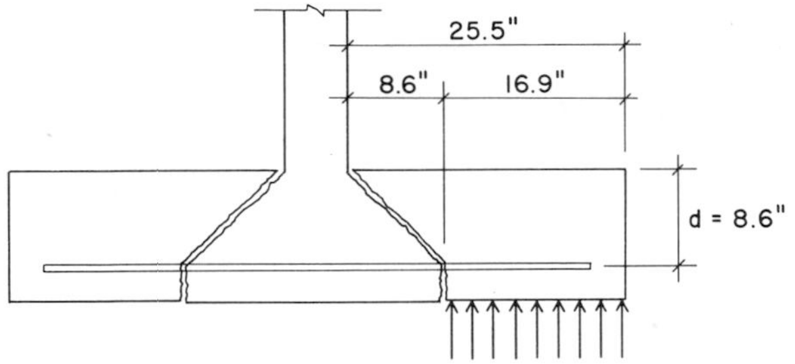

FIGURE 3.19. Shear action in the wall footing–Example 2.

the wall. As shown in Figure 3.19, this moves the shear section out to 16.9 in. from the edge of the footing. This is reasonably valid when the cantilever distance is significantly larger than the depth of the section, but is questionable for short cantilevers. In fact, the code recommends that this shortened span not be used for brackets and short cantilevers. We therefore recommend that the critical section for shear be taken at the face of the wall, unless the cantilever exceeds three times the footing thickness. However, if the latter assumption is made, it is reasonable to use the full thickness of the footing, rather than the effective *d* distance of the cross section. We will illustrate both calculations for our example.

Case 1–shear at the *d* distance from the wall (See Figure 3.20.)

Shear force:

$$V = (1842)\left(\frac{16.9}{12}\right) = 2594 \text{ lb}$$

Stress:

$$v_c = \frac{V}{bd} = \frac{2594}{(12)(8.6)} = 25 \text{ psi}$$

Case 2–shear at the wall face:

Shear force:

$$V = (1842)\left(\frac{25.5}{12}\right) = 3914 \text{ lb}$$

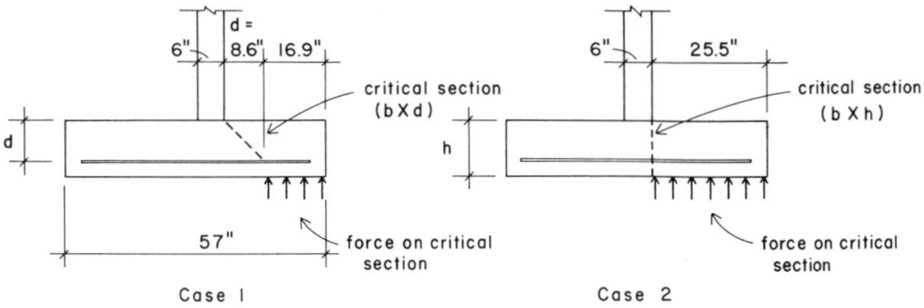

FIGURE 3.20. Comparison of shear analyses for a long cantilever (Case 1) and a short cantilever (Case 2).

Stress:

$$v_c = \frac{V}{bh} = \frac{3914}{(12)(12)} = 27 \text{ psi}$$

Both of these are well below the allowable stress of $1.1\sqrt{f_c'}$, which is the same as in the previous example: 49 psi. It is possible, therefore, to reduce the footing thickness if the shear stress is considered to be an important criterion. However, as has been discussed previously, reduction of cost in construction is usually obtained by minimizing the amount of reinforcing, and any reduction in the footing thickness will shorten the moment arm for the tension reinforcing, requiring an increase in steel area. It therefore becomes a matter of judgment about the ideal value for the footing thickness.

If we reduce the footing thickness to 11 in., a second try would proceed as follows:

$$\text{new footing weight} = \tfrac{11}{12}(150) = 137.5, \text{ say } 138 \text{ lb/ft}^2$$

$$\text{usable soil pressure } (p) = 2000 - 138 = 1862 \text{ lb/ft}^2$$

$$\text{required width } (w) = \frac{8750}{1862} = 4.70 \text{ ft or } 56.4 \text{ in.}$$

which does not change the footing width or design soil pressure.

$$\text{new } d = h - 3 - \frac{D}{2} = 11 - 3.375 = 7.625, \text{ say } 7.6 \text{ in.}$$

For the Case 2 shear stress, the shear force is the same as for the first try, and the new shear stress is

$$v_c = \frac{V}{bh} = \frac{3914}{(12)(11)} = 30 \text{ psi}$$

For the Case 1 shear stress, the shear section is now an inch closer to the wall and the shear force becomes

$$V = (1842)\left(\frac{17.9}{12}\right) = 2748 \text{ lb}$$

$$v_c = \frac{V}{bd} = \frac{2748}{(12)(7.6)} = 30 \text{ psi}$$

The bending moment to be used for concrete stress and determination of the steel area is

$$M = (3914)\left(\frac{25.5}{2}\right) = 49{,}903 \text{ lb-in.}$$

and the required steel area per foot of wall length is

$$A_s = \frac{M}{f_s jd} = \frac{49{,}903}{(20)(0.9)(7.6)} = 0.365 \text{ in.}^2$$

Bar sizes and spacings can be selected most easily with the use of handbook tables giving the average steel areas for various combinations of bar size and spacing. Examples of such tables may be found in the *ACI* working stress handbook (Ref. 7) or the *CRSI* handbook (Ref. 6). If no table is available, possible spacings may be calculated as follows:

$$\text{area of steel/ft} = (\text{area of one bar}) \; \frac{12}{\text{spacing of bars (in.)}}$$

thus

$$\text{required spacing} = (\text{area of one bar}) \; \frac{12}{\text{Required area/ft}}$$

Using this relationship, the required spacings for bar sizes 3 through 6 are shown in Table 3.3. Selection of the actual bar size and spacing

TABLE 3.3. Selection of Reinforcing for Example 2

Bar Size	Area of Bar (in.^{2})	Area Required for Flexure (in.^{2})	Spacing Required (in.)	Selected Spacing (in.)
3	0.11	0.365	3.6	$3\frac{1}{2}$
4	0.20	0.365	6.6	$6\frac{1}{2}$
5	0.31	0.365	10.2	10
6	0.44	0.365	14.5	$14\frac{1}{2}$
7	0.60	0.365	19.7	$19\frac{1}{2}$

is a matter of design judgment for which we provide the following considerations.

1. Maximum recommended spacing is 18 in.
2. Minimum recommended spacing is 6 in. to minimize the number of bars and allow for easy placing of concrete.
3. Preference is for smaller bars as long as spacing is not too close.

With these considerations in mind our selection would be for the No. 5 bars at 10 in. center-to-center spacing, although the No. 4 or No. 6 bars would also be feasible.

Flexural stress, of course, is only one consideration. The other problem is that of bond stress on the surface of the bars. This may be dealt with in one of two ways. The first way is to calculate a perimeter requirement for the steel similar to the area requirement for flexure, and then to select the bar spacing to satisfy both requirements. The problem with this method is that the allowable bond stress is a function of the bar diameter and thus cannot be calculated on a single-value basis as the area can. From the 1963 ACI code (Ref. 11) the allowable bond stress is stated as

$$u = \frac{4.8\sqrt{f_c'}}{D} \quad \text{or a maximum of 500 psi}$$

Values for the allowable bond stress calculated from this limitation are given in the appendix (Table A.2) for various values of f_c'. As may be seen from the table, the 500 psi limit applies to the smaller bar sizes, whereas the formula provides the limit for larger bars.

The second possible procedure for dealing with bond stress is simply to calculate the area required for flexure, select the bars for the area, and check the bond stress on the selected bars. Then if bond stress is shown to be critical, a revision can be made to select smaller bars or revise the spacing.

For our calculation the critical shear force to be used for the bond stress calculation is the same as that used for the cantilever moment: 3914 lb. The total perimeter of the bars for our selection of No. 5 bars at 10 in. is determined as follows:

$$\text{avg. } \Sigma_0/\text{ft} = (\text{perimeter of one bar})\left(\tfrac{12}{10}\right)$$
$$= (1.963)\left(\tfrac{12}{10}\right) = 2.356 \text{ in./ft}$$

and the bond stress is determined as

$$u = \frac{V}{\Sigma_0 jd} = \frac{3914}{(2.356)(0.9)(7.6)} = 243 \text{ psi}$$

Using the code formula the allowable stress is determined to be

$$u = \frac{4.8\sqrt{f_c'}}{D} = \frac{4.8\sqrt{2000}}{0.625} = 343 \text{ psi}$$

so that our choice is adequate for both stress conditions.

As with the unreinforced footing we recommend a minimum reinforcing in the long direction of 0.0015 times the cross section of the footing, which is an area of

$$A_s = (0.0015)(11)(57) = 0.94 \text{ in.}^2$$

Using three No. 5 bars,

$$A_s = (3)(0.31) = 0.93 \text{ in.}^2$$

3.3 Square Column Footings

The ordinary footing used under concentrated loads such as those developed in columns is a square pad of concrete. Although a round pad would be more logical structurally, the practical aspects of forming and reinforcing make the square form more reasonable. Round or polygonal forms are sometimes used under large towers, but almost all foundations for building columns are rectangular.

As with the wall footing, the two concrete stresses of concern are tension due to bending and diagonal tension due to shear. For the wall footing the bending action was in a single direction. For the column footing the bending action is literally in all directions, although design procedures for the rectangular footing consist of designing the footing for one-way bending with the same moment and shear effects as those in the wall footing. The symmetrical footing is then made to work for this bending in the two mutually perpendicular directions.

For the column footing, however, a second shear analysis is made on the basis of so-called punching shear. The diagonal tension failure in this case is visualized in the form of a pyramid shape around the column edge, as shown in Figure 3.21. The shear face assumed for this is taken at a distance out from the column of one half of the effective depth for a reinforced footing.

As with the wall footing, it is possible to use an unreinforced footing as long as stresses in the concrete are low. Table 3.4 gives sizes for unreinforced column footings for a range of soil pressures and two concrete strengths. The maximum size recommended for such a footing is 3 ft square, which is the largest size listed in the table. The following example illustrates the basis for determination of the entries in Table 3.4.

Example 3 Unreinforced Column Footing

Design data and criteria:

Column load: 15,000 lb.
Column size: 8 in. square.
Maximum allowable soil pressure: 2000 lb/ft^2.
Concrete design strength: $f_c' = 2000$ psi.

The 8 in. size is somewhat small for a concrete column since this is less than the minimum size allowed for such a column. However, the codes require that the size used for design with a masonry column be one half the actual size of the column. The 8 in. design size therefore relates to a 16 in. masonry column, which is a typical minimum size. For a footing under a steel column the effective column size for design is taken at

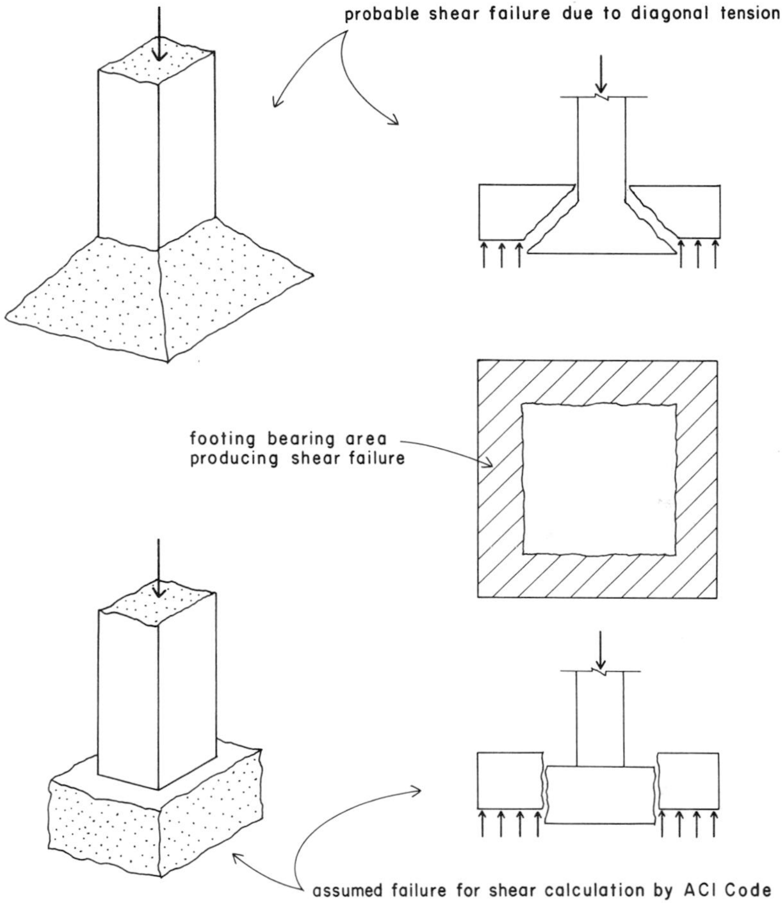

FIGURE 3.21. Punching shear action in column footings.

half the distance between the face of the column and the edge of the base plate.

The design procedure is essentially similar to that for the wall footing. A footing thickness is assumed, a width is found from the total soil pressure, the footing is checked for stresses to verify the thickness assumption, and revision is made if necessary.

TABLE 3.4. Allowable Loads on Unreinforced Square Column Footings[a]

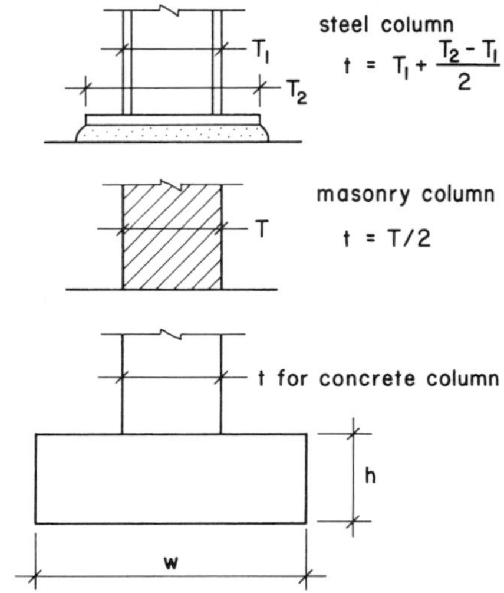

[a] For illustration and discussion of the process used to generate the table data see Example 3 in the text. Allowable loads do not include the weight of the footing, which has been deducted from the total bearing capacity. Example: 1000 lb/ft² soil pressure, 6 in. thick by 30 in. square footing; total capacity = $(1000)(2.5)^2 = 6250$ lb; footing weighs 75 lb/ft²; allowable load = $(1000 - 75)(2.5)^2 = 5781$ lb.

Maximum Soil Pressure (lb/ft²)	Minimum Column Width t (in.)	$f'_c = 2000$ psi			$f'_c = 3000$ psi		
		Allowable Load on Footing (lb)	Footing Dimensions h (in.)	Footing Dimensions w (in.)	Allowable Load on Footing (lb)	Footing Dimensions h (in.)	Footing Dimensions w (in.)

1000	4	2081	6	18	2081	6	18
	4	2833	6	21	2833	6	21
	4	3700	6	24	3700	6	24
	4	4683	6	27	4683	6	27
	4	5703	7	30	5703	7	30
	8	5781	6	30	5781	6	30
	4	6806	8	33	6900	7	33
	8	6900	7	33	6995	6	33
	4	7988	9	36	8100	8	36
	8	8100	8	36	8212	7	36
1500	4	3206	6	18	3206	6	18
	4	4364	6	21	4364	6	21
	4	5650	7	24	5700	6	24
	8	5700	6	24	5700	6	24
	4	7088	8	27	7148	7	27
	8	7148	7	27	7214	6	27
	4	8672	9	30	8750	8	30
	8	8750	8	30	8828	7	30
	4	10398	10	33	10493	9	33
	8	10588	8	33	10588	8	33
	4	12262	11	36	12375	10	36
	8	12488	9	36	12488	9	36
2000	4	4331	6	18	4331	6	18
	4	5857	7	21	5895	6	21
	8	5895	6	21	5895	6	21
	4	7600	8	24	7650	7	24
	8	7700	6	24	7700	6	24
	4	9555	9	27	9619	8	27
	8	9619	8	27	9682	7	27

TABLE 3.4. (*Continued*)

Maximum Soil Pressure (lb/ft^2)	Minimum Column Width t (in.)	f'_c = 2000 psi			f'_c = 3000 psi		
		Allowable Load on Footing (lb)	Footing Dimensions h (in.)	Footing Dimensions w (in.)	Allowable Load on Footing (lb)	Footing Dimensions h (in.)	Footing Dimensions w (in.)
	4	11719	10	30	11797	9	30
	8	11797	9	30	11875	8	30
	4	14085	11	33	14180	10	33
	8	14180	10	33	14274	9	33
	4	16650	12	36	16762	11	36
	8	16762	11	36	16875	10	36
3000	4	6553	7	18	6581	6	18
	8	6581	6	18	6581	6	18
	4	8881	8	21	8881	8	21
	8	8958	6	21	8958	6	21
	4	11500	10	24	11550	9	24
	8	11600	8	24	11650	7	24
	4	14491	11	27	14555	10	27
	8	14618	9	27	14681	8	27
	4	17734	13	30	17813	12	30
	8	17891	11	30	17969	10	30
	4	21270	15	33	21364	14	33
	8	21553	12	33	21648	11	33
	4	24975	18	36	25200	16	36
	8	25538	13	36	25650	12	36

Try: $h = 11$ in.

Then: footing weight $= \frac{11}{12}(150) = 137.5$, say 138 lb/ft^2

net usable soil pressure $(p) = 2000 - 138 = 1862$ lb/ft^2

required area $(A) = \dfrac{15{,}000}{1862} = 8.05$ ft^2

required width $(w) = \sqrt{8.05} = 2.84$ ft or 34.05 in.

Try: $w = 34$ in.

Then: design soil pressure $= \dfrac{15{,}000}{(34/12)^2} = 1869$ lb/ft^2

As shown in Figure 3.22, bending is produced by the soil pressure on an area 34 by 13 in., and the bending section is 34 in. wide and 11 in. high.

Bending force:

$$F = (1869)\left(\tfrac{34}{12}\right)\left(\tfrac{13}{12}\right) = 5737 \text{ lb}$$

Moment:

$$M = (5737)\left(\tfrac{13}{2}\right) = 37{,}289 \text{ lb-in.}$$

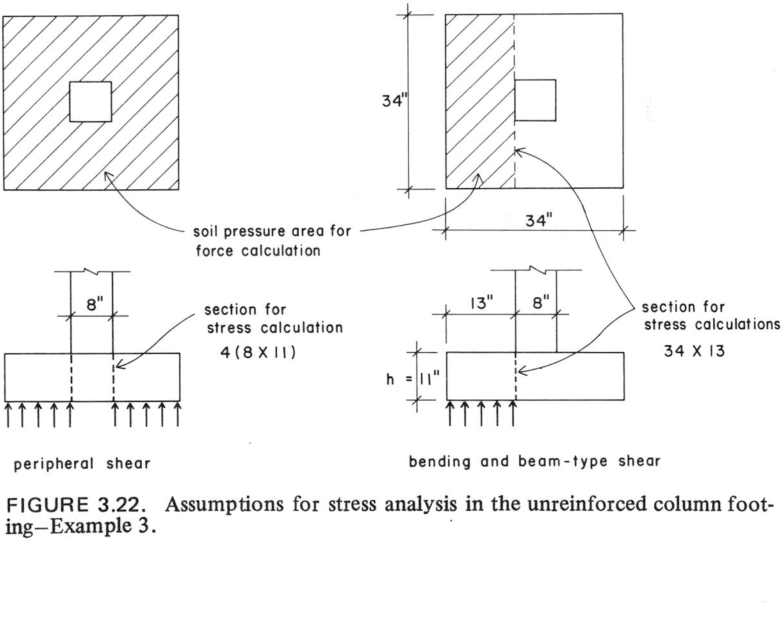

FIGURE 3.22. Assumptions for stress analysis in the unreinforced column footing–Example 3.

Section modulus:

$$S = \frac{(34)(11)^2}{6} = 686 \text{ in.}^3$$

Bending stress:

$$f_c = \frac{M}{S} = \frac{37{,}289}{686} = 54 \text{ psi}$$

This stress is compared to the allowable stress given by the ACI code as

$$f_c = 1.6\sqrt{f_c'} = 1.6\sqrt{2000} = 72 \text{ psi}$$

For the unreinforced footing we make a conservative shear analysis by assuming the shear face to be at the face of the column. For the beam-type shear, as shown in Figure 3.22, the shear force is thus the same as the bending force, and the shear section is the same as the bending section. Thus the shear stress is

$$v_c = \frac{V}{wh} = \frac{5737}{(34)(11)} = 15 \text{ psi}$$

For the peripheral shear, the shear force is the full column load reduced by the soil pressure below the column area. The shear force is thus

$$V = 15{,}000 - (1869)(\tfrac{8}{12})^2 = 15{,}000 - 831 = 14{,}169 \text{ lb}$$

The total peripheral shear face, as shown in Figure 3.22, is the sum of the four faces around the column, each 11 by 8 in. The peripheral shear stress is thus

$$v_c = \frac{14{,}169}{(4)(11 \times 8)} = 40 \text{ psi}$$

The allowable stresses for these two conditions are as follows:

For beam shear:

$$v_c = 1.1\sqrt{f_c'} = 1.1\sqrt{2000} = 49 \text{ psi}$$

For peripheral shear:

$$v_c = 2\sqrt{f_c'} = 2\sqrt{2000} = 89 \text{ psi}$$

Since all three calculated stresses are well below the limits, it is possible to reduce the footing thickness. Reducing the thickness to 10 in. will result in the following:

$$\text{new footing weight} = \tfrac{10}{12}\,(150) = 125 \text{ lb/ft}^2$$

$$\text{usable soil pressure } (p) = 2000 - 125 = 1875 \text{ lb/ft}^2$$

$$\text{area required } (A) = \frac{15{,}000}{1875} = 8.0 \text{ ft}^2$$

$$\text{required width } (w) = \sqrt{8.00} = 2.83 \text{ ft or } 33.9 \text{ in.}$$

Using the same width, soil pressure, bending force and moment

$$S = \frac{(34)(10)^2}{6} = 567 \text{ in.}^3$$

$$f_c = \frac{37{,}289}{567} = 66 \text{ psi (less than 72)}$$

Shear stresses will increase only slightly. Unless fractional inch dimensions are to be used, this is the limit based on the bending stress allowed.

As in the case of the wall footings without transverse reinforcing, it is advisable to increase the calculated thickness when the footing is poured on dry, porous soil.

Some designers prefer to use reinforcing in column footings as small as 2 ft square. There is nothing wrong with this, although it can usually be shown that the footing is adequate without reinforcing in most cases up to a width of 3 ft or so, as long as soil pressures are reasonably low. For larger footings and higher soil pressures it becomes necessary to use reinforcing to reduce the footing mass. For exceptionally large footings it may be necessary to use some additional measures for reduction of the footing mass, involving the use of piers, stiffening ribs, and so on. These extraordinary measures are discussed in Section 3.6.

Table 3.5 gives data for predesigned footings for two concrete strengths and several soil pressures. The following example illustrates the basis for determination of the table entries.

TABLE 3.5. Allowable Loads on Reinforced Square Column Footings[a]

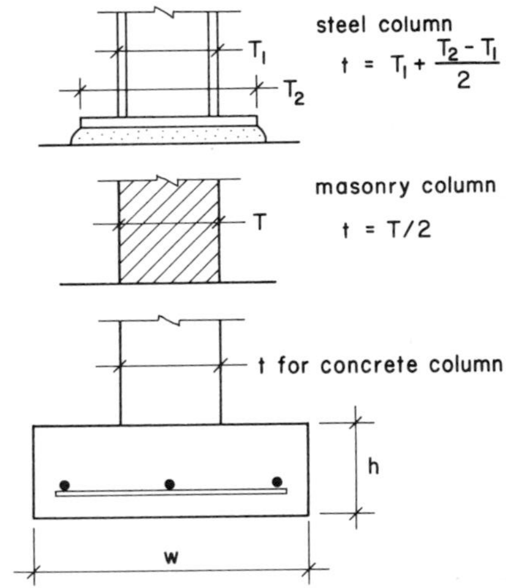

[a] For illustration and discussion of the process used to generate the table data see Example 4 in the text. Allowable loads do not include the weight of the footing, which has been deducted from the total bearing capacity. Example: 1000 lb/ft^2 soil pressure, 12 in. thick by 7 ft square footing; total capacity = $(1.00)(7)^2$ = 49 k; footing weighs 150 lb/ft^2; allowable load = $(1.00 - 0.15)(7)^2$ = 41.65 k; round off to 42 k.

		f'_c = 2000 psi				f'_c = 3000 psi			
			Footing Dimensions				Footing Dimensions		
Maximum Soil Pressure (lb/ft^2)	Minimum Column Width t (in.)	Allowable Load on Footing (k)	h (in.)	w (ft)	Reinforcing Each Way	Allowable Load on Footing (k)	h (in.)	w (ft)	Reinforcing Each Way

1000	8	5.5	10	2.5	2 No. 3	5.5	10	2.5	2 No. 3
	8	7.9	10	3.0	2 No. 3	7.9	10	3.0	2 No. 3
	8	10.7	10	3.5	3 No. 3	10.7	10	3.5	3 No. 3
	8	14.0	10	4.0	3 No. 4	14.0	10	4.0	3 No. 4
	8	17.7	10	4.5	4 No. 4	17.7	10	4.5	4 No. 4
	8	22	10	5.0	4 No. 5	22	10	5.0	4 No. 5
	8	26	10	5.5	5 No. 5	26	10	5.5	5 No. 5
	8	31	10	6.0	5 No. 6	31	10	6.0	5 No. 6
	8	36	11	6.5	5 No. 6	37	10	6.5	6 No. 6
	8	42	12	7.0	6 No. 6	42	11	7.0	7 No. 6
	8	48	12	7.5	7 No. 6	48	12	7.5	7 No. 6
1500	8	8.6	10	2.5	2 No. 3	8.6	10	2.5	2 No. 3
	8	12.4	10	3.0	3 No. 3	12.4	10	3.0	3 No. 3
	8	16.8	10	3.5	3 No. 4	16.8	10	3.5	3 No. 4
	8	22	10	4.0	4 No. 4	22	10	4.0	4 No. 4
	8	28	10	4.5	4 No. 5	28	10	4.5	4 No. 5
	8	34	11	5.0	5 No. 5	34	10	5.0	6 No. 5
	8	41	11	5.5	7 No. 5	41	11	5.5	7 No. 5
	8	48	12	6.0	6 No. 6	49	11	6.0	6 No. 6
	8	56	13	6.5	6 No. 6	57	12	6.5	7 No. 6
	8	65	14	7.0	7 No. 6	65	13	7.0	6 No. 7
	8	74	15	7.5	6 No. 7	74	14	7.5	7 No. 7
	8	83	16	8.0	7 No. 7	84	15	8.0	7 No. 7
	8	93	17	8.5	8 No. 7	94	16	8.5	8 No. 7
	8	103	18	9.0	8 No. 7	105	16	9.0	10 No. 7
	8	115	18	9.5	10 No. 7	116	17	9.5	10 No. 7
	10	126	19	10.0	10 No. 7	129	17	10.0	9 No. 8

TABLE 3.5. (*Continued*)

Maximum Soil Pressure (lb/ft²)	Minimum Column Width t (in.)	$f'_c = 2000$ psi				$f'_c = 3000$ psi			
		Allowable Load on Footing (k)	Footing Dimensions h (in.)	Footing Dimensions w (ft)	Reinforcing Each Way	Allowable Load on Footing (k)	Footing Dimensions h (in.)	Footing Dimensions w (ft)	Reinforcing Each Way
2000	8	12	10	2.5	2 No. 3	12	10	2.5	2 No. 3
	8	17	10	3.0	4 No. 3	17	10	3.0	4 No. 3
	8	23	10	3.5	4 No. 4	23	10	3.5	4 No. 4
	8	30	10	4.0	6 No. 4	30	10	4.0	6 No. 4
	8	37	11	4.5	5 No. 5	38	10	4.5	6 No. 5
	8	46	12	5.0	6 No. 5	46	11	5.0	5 No. 6
	8	55	13	5.5	5 No. 6	56	12	5.5	6 No. 6
	8	65	14	6.0	6 No. 6	66	13	6.0	7 No. 6
	8	76	15	6.5	7 No. 6	77	14	6.5	6 No. 7
	8	88	16	7.0	8 No. 6	89	15	7.0	7 No. 7
	8	100	17	7.5	7 No. 7	101	16	7.5	8 No. 7
	8	113	18	8.0	8 No. 7	114	17	8.0	9 No. 7
	8	127	19	8.5	9 No. 7	128	18	8.5	7 No. 8
	8	142	20	9.0	8 No. 8	143	19	9.0	8 No. 8
	8	157	21	9.5	8 No. 8	158	20	9.5	9 No. 8
	10	174	21	10.0	9 No. 8	175	20	10.0	10 No. 8
	10	189	23	10.5	10 No. 8	191	21	10.5	11 No. 8

	10	207	23	11.0	11 No. 8	208	22	11.0	12 No. 8
	10	225	24	11.5	12 No. 8	226	23	11.5	10 No. 9
	10	243	25	12.0	13 No. 8	245	24	12.0	11 No. 9
3000	8	18	10	2.5	2 No. 4	18	10	2.5	2 No. 4
	8	26	10	3.0	3 No. 4	26	10	3.0	3 No. 4
	8	35	10	3.5	4 No. 5	35	10	3.5	4 No. 5
	8	45	12	4.0	4 No. 5	46	11	4.0	5 No. 5
	8	57	13	4.5	6 No. 5	57	12	4.5	6 No. 5
	8	70	14	5.0	5 No. 6	71	13	5.0	6 No. 6
	8	84	16	5.5	6 No. 6	85	14	5.5	7 No. 6
	8	100	17	6.0	7 No. 6	101	15	6.0	8 No. 6
	8	117	18	6.5	9 No. 6	118	17	6.5	7 No. 7
	10	135	19	7.0	7 No. 7	136	18	7.0	8 No. 7
	10	154	20	7.5	8 No. 7	156	18	7.5	7 No. 8
	10	175	21	8.0	10 No. 7	177	19	8.0	8 No. 8
	10	197	22	8.5	11 No. 7	198	21	8.5	9 No. 8
	12	219	23	9.0	9 No. 8	221	21	9.0	10 No. 8
	12	243	24	9.5	10 No. 8	246	22	9.5	9 No. 9
	12	269	25	10.0	11 No. 8	271	23	10.0	10 No. 9
	12	293	27	10.5	10 No. 9	297	24	10.5	11 No. 9
	12	320	28	11.0	11 No. 9	323	26	11.0	12 No. 9
	12	349	29	11.5	12 No. 9	352	27	11.5	10 No. 10
	14	378	30	12.0	12 No. 9	381	28	12.0	11 No. 10
	14	408	31	12.5	13 No. 9	412	29	12.5	12 No. 10
	14	439	32	13.0	12 No. 10	443	30	13.0	13 No. 10
	14	471	33	13.5	13 No. 10	476	31	13.5	14 No. 10

TABLE 3.5. *(Continued)*

		f'_c = 2000 psi				f'_c = 3000 psi			
			Footing Dimensions				Footing Dimensions		
Maximum Soil Pressure (lb/ft^2)	Minimum Column Width t (in.)	Allowable Load on Footing (k)	h (in.)	w (ft)	Reinforcing Each Way	Allowable Load on Footing (k)	h (in.)	w (ft)	Reinforcing Each Way
	14	504	35	14.0	13 No. 10	509	32	14.0	15 No. 10
	16	539	35	14.5	14 No. 10	546	32	14.5	16 No. 10
	16	574	36	15.0	15 No. 10	582	33	15.0	17 No. 10
4000	8	24	10	2.5	3 No. 4	24	10	2.5	3 No. 4
	8	35	10	3.0	4 No. 4	35	10	3.0	4 No. 4
	8	47	12	3.5	4 No. 5	47	11	3.5	4 No. 5
	8	61	13	4.0	5 No. 5	61	12	4.0	6 No. 5
	8	77	15	4.5	5 No. 6	77	13	4.5	6 No. 6
	8	95	16	5.0	6 No. 6	95	15	5.0	6 No. 6
	8	113	18	5.5	7 No. 6	115	16	5.5	6 No. 7
	8	135	19	6.0	8 No. 6	136	18	6.0	7 No. 7
	8	158	21	6.5	7 No. 7	159	19	6.5	8 No. 7
	10	182	22	7.0	8 No. 7	184	20	7.0	9 No. 7
	10	209	23	7.5	9 No. 7	210	21	7.5	8 No. 8
	10	237	24	8.0	9 No. 8	238	22	8.0	9 No. 8
	10	265	26	8.5	9 No. 8	267	24	8.5	10 No. 8
	12	297	26	9.0	10 No. 8	299	24	9.0	9 No. 9

	12	329	28	9.5	11 No. 8	331	26	9.5	10 No. 9
	12	364	29	10.0	13 No. 8	366	27	10.0	11 No. 9
	12	398	31	10.5	11 No. 9	402	28	10.5	12 No. 9
	14	435	32	11.0	12 No. 9	440	29	11.0	11 No. 10
	14	474	33	11.5	13 No. 9	479	30	11.5	12 No. 10
	14	515	34	12.0	14 No. 9	520	31	12.0	13 No. 10
	14	554	36	12.5	15 No. 9	560	33	12.5	14 No. 10
	16	600	36	13.0	17 No. 9	606	33	13.0	15 No. 10
	16	642	38	13.5	18 No. 9	651	34	13.5	16 No. 10
	16	688	39	14.0	15 No. 10	696	36	14.0	14 No. 11
	16	733	41	14.5	16 No. 10	744	37	14.5	15 No. 11
	18	784	41	15.0	17 No. 10	793	38	15.0	16 No. 11
	18	832	43	15.5	18 No. 10	844	39	15.5	17 No. 11
	18	883	44	16.0	20 No. 10	896	40	16.0	18 No. 11
	18	936	46	16.5	17 No. 11	949	41	16.5	19 No. 11
	20	990	46	17.0	18 No. 11	1004	42	17.0	20 No. 11
	20	1045	47	17.5	19 No. 11	1060	43	17.5	21 No. 11
	20	1097	49	18.0	20 No. 11	1118	44	18.0	23 No. 11
5000	8	30	10	2.5	3 No. 4	30	10	2.5	3 No. 4
	8	44	11	3.0	5 No. 4	44	10	3.0	4 No. 5
	8	59	13	3.5	6 No. 4	59	12	3.5	5 No. 5
	8	77	14	4.0	6 No. 5	77	13	4.0	5 No. 6
	8	97	16	4.5	7 No. 5	97	15	4.5	6 No. 6
	8	119	18	5.0	6 No. 6	120	16	5.0	7 No. 6
	8	144	19	5.5	8 No. 6	144	18	5.5	9 No. 6
	10	171	20	6.0	9 No. 6	171	19	6.0	7 No. 7

TABLE 3.5. (*Continued*)

Maximum Soil Pressure (lb/ft²)	Minimum Column Width t (in.)	$f'_c = 2000$ psi: Allowable Load on Footing (k)	Footing Dimensions h (in.)	Footing Dimensions w (ft)	Reinforcing Each Way	$f'_c = 3000$ psi: Allowable Load on Footing (k)	Footing Dimensions h (in.)	Footing Dimensions w (ft)	Reinforcing Each Way
	10	199	22	6.5	8 No. 7	200	20	6.5	9 No. 7
	10	230	24	7.0	9 No. 7	231	22	7.0	8 No. 8
	10	263	25	7.5	11 No. 7	265	23	7.5	9 No. 8
	12	299	26	8.0	12 No. 7	301	24	8.0	10 No. 8
	12	336	28	8.5	13 No. 7	338	25	8.5	12 No. 8
	12	374	30	9.0	11 No. 8	377	27	9.0	10 No. 9
	12	416	31	9.5	13 No. 8	419	28	9.5	12 No. 9
	14	460	32	10.0	14 No. 8	464	29	10.0	13 No. 9
	14	504	34	10.5	15 No. 8	510	31	10.5	13 No. 9
	14	552	35	11.0	17 No. 8	556	32	11.0	15 No. 9
	14	600	37	11.5	15 No. 9	605	34	11.5	13 No. 10
	16	651	38	12.0	16 No. 9	657	35	12.0	14 No. 10
	16	705	39	12.5	17 No. 9	711	36	12.5	15 No. 10
	16	758	41	13.0	18 No. 9	767	37	13.0	16 No. 10
	18	815	42	13.5	16 No. 10	824	38	13.5	18 No. 10
	18	874	43	14.0	17 No. 10	884	39	14.0	16 No. 11
	18	933	45	14.5	18 No. 10	943	41	14.5	16 No. 11
	20	995	46	15.0	19 No. 10	1007	42	15.0	17 No. 11

20	1060	47	15.5	20 No. 10	1072	43	15.5	19 No. 11
20	1123	49	16.0	22 No. 10	1136	45	16.0	20 No. 11
22	1191	50	16.5	19 No. 11	1204	46	16.5	21 No. 11
22	1261	51	17.0	20 No. 11	1275	47	17.0	22 No. 11
22	1328	53	17.5	21 No. 11	1347	48	17.5	24 No. 11
24	1401	54	18.0	22 No. 11	1421	49	18.0	25 No. 11
24	1476	55	18.5	23 No. 11	1497	50	18.5	26 No. 11
26	1552	56	19.0	24 No. 11	1575	51	19.0	27 No. 11
26	1630	57	19.5	26 No. 11	1654	52	19.5	29 No. 11
28	1710	58	20.0	27 No. 11	1735	53	20.0	30 No. 11
28	1791	59	20.5	29 No. 11	1817	54	20.5	32 No. 11
30	1874	60	21.0	30 No. 11	1902	55	21.0	33 No. 11
30	1959	61	21.5	31 No. 11	1987	56	21.5	35 No. 11
32	2045	62	22.0	33 No. 11	2075	57	22.0	36 No. 11

EXAMPLE 4 REINFORCED SQUARE COLUMN FOOTING

Design data and criteria:

Column load: 500 k.
Column size: 15 in. square.
Maximum allowable soil pressure: 4000 lb/ft^2.
Concrete design strength: $f'_c = 3000$ psi.
Allowable tension on reinforcing: 20,000 psi.

For this large a footing the first guess for the footing thickness is a real shot in the dark. Since we have demonstrated the process of guessing and revising in previous examples, we will not do so here.

Try: $h = 30$ in.
Then: footing weight $= \frac{30}{12}(150) = 375$ lb/ft^2
net usable soil pressure $= 4000 - 375 = 3625$ lb/ft^2

$$\text{required area} = \frac{500{,}000}{3625} = 137.9 \text{ ft}^2$$

required width $= \sqrt{137.9} = 11.74$ ft

Try: $w = 11$ ft 9 in. or 11.75 ft

$$\text{Then: design soil pressure} = \frac{500{,}000}{(11.75)^2} = 3{,}622 \text{ lb/ft}^2$$

Determination of the bending force and bending moment are as shown in Figure 3.23.

Bending force:

$$F = (3622)(\tfrac{63}{12})(11.75) = 223{,}432 \text{ lb}$$

Moment:

$$M = (223{,}432)(\tfrac{63}{12})(\tfrac{1}{2}) = 586{,}509 \text{ lb-ft}$$

This moment is assumed to operate in both directions on the footing and the footing is typically reinforced with the same number and size of bars in each direction. However, it is necessary to place the bars in one direction on top of the bars in the other direction, as shown in

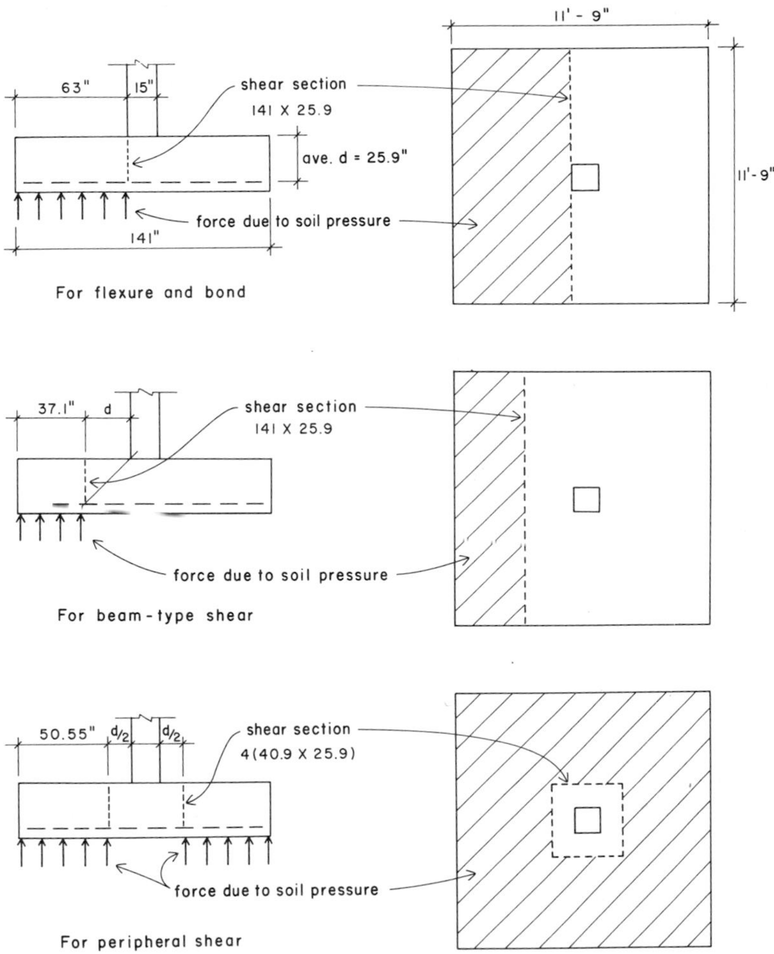

FIGURE 3.23. Assumptions for stress analysis in the reinforced column footing–Example 4.

Figure 3.24, and there are thus different effective depths in the two directions. For the stress calculations we recommend the use of an average depth, which results in a slight overstress in one direction compensated by a slight understress in the other direction. It is also necessary to assume a size for the reinforcing in order to determine the effective depth. As with the footing thickness, this must be a guess,

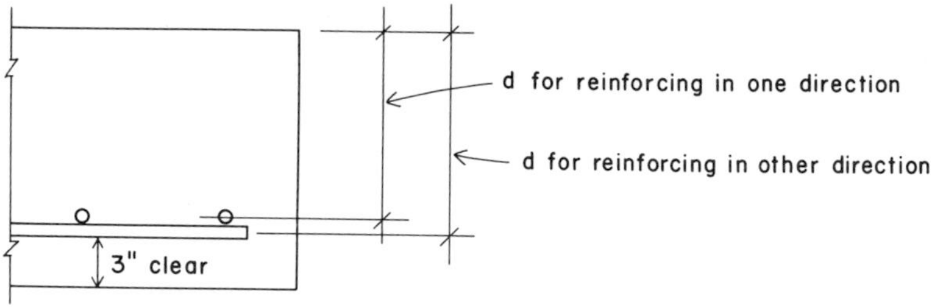

FIGURE 3.24. Effective depth for the two-way-reinforced footing.

unless some reference is used. Assuming a No. 9 bar for our footing, the effective depth thus becomes

$$\text{average } d = h - 3 - (\text{bar } D) = 30 - 3 - 1.13 = 25.87 \text{ in., say } 25.9 \text{ in.}$$

The section resisting the bending moment is therefore one that is 141 in. wide and has a d of 25.9 in. The balanced moment capacity of this section is

$$M = Rbd^2 = (226)(141)(25.9)^2(\tfrac{1}{12}) = 1{,}781{,}336 \text{ lb-ft}$$

which is more than three times the required moment.

From this analysis it may be seen that the compressive bending stress in the concrete will not be critical. Furthermore, the values for k and j will be those for a considerably underreinforced section. Referring to Table A-1 in the appendix, the value for a balanced k is 0.383. A conservative assumption for the true k value and the corresponding j value is

$$k = 0.30,\ j = 1 - \frac{k}{3} = 1 - \frac{0.30}{3} = 0.90$$

The critical stress condition in the concrete is that of shear, either in beam-type action or punching action. Referring to the illustrations in Figure 3.23, the analysis for these two conditions is as follows:

Beam-type shear:

Shear force:

$$V = (3622)(11.75)\left(\frac{37.1}{12}\right) = 131{,}577 \text{ lb}$$

Shear stress:

$$v_c = \frac{131,577}{(141)(25.9)} = 36 \text{ psi}$$

Allowable stress:

$$v_c = 1.1\sqrt{f_c'} = 1.1\sqrt{3000} = 60 \text{ psi}$$

Peripheral shear:

Shear force:

$$V = (3622)\left[(11.75)^2 - \left(\frac{40.9}{12}\right)^2\right]$$

$$= (3622)(138 - 11.6)$$

$$= 457,821 \text{ lb}$$

Shear stress:

$$v_c = \frac{457,821}{(4)(40.9)(25.9)} = 108 \text{ psi}$$

Allowable stress:

$$v_c = 2\sqrt{f_c'} = 2\sqrt{3000} = 110 \text{ psi}$$

Although the beam shear is low, the peripheral stress is just barely short of the limit, so the 30 in. thickness is indeed the minimum allowable dimension.

Using the assumed value of 0.9 for j, the area of steel required is determined as

$$A_s = \frac{M}{f_s jd} = \frac{(586,509)(12)}{(20,000)(0.9)(25.9)} = 15.1 \text{ in.}^2$$

There are a number of bar size-and-number combinations that may be selected to satisfy this area requirement. Another stress condition that must be considered is the bond on the surface of the bars. For a specific situation there will be a maximum bar size that can be used if the full potential of the area is to be usable in tension. This is because the bond stress is limited by the bar size, as well as the concrete strength. Table 3.6 lists a range of bar sizes that may be used to satisfy the area require-

TABLE 3.6. Possible Bar Combinations (Example 4)

Bars (Number and Size)	Actual A_s Provided (in.^{2})	Total Perimeter Σ_0 (in.)	Actual Bond Stress $u = \dfrac{V}{\Sigma_0 jd}$ (psi)	Allowable Bond Stress $u = \dfrac{4.8\sqrt{f'_c}}{D}$ (psi)	Approximate Center-to-Center Spacing (in.)
26 No. 7	15.60	71.5	134	300	5.3
20 No. 8	15.80	62.8	153	263	7.0
16 No. 9	16.00	56.7	169	233	8.9
12 No. 10	15.24	47.9	200	207	12.1
10 No. 11	15.60	44.3	216	186	14.8

ment just determined. Also shown in the table are the calculation of actual bond stress on the bars and the allowable bond stress as determined by the code formula (see the appendix). The bond stresses are found as follows:

Actual stress:

$$u = \frac{V}{\Sigma_0 jd}$$

where V is the bending force: 223,432 lb, and

Allowable stress:

$$u = \frac{4.8\sqrt{f'_c}}{D}$$

where D is the bar diameter.

From the calculations in the table it may be seen that the No. 10 bars are the largest in which the cross sectional area can be fully developed in tension. The other consideration is the spacing of the bars. For 12 No. 10 bars, assuming the first bar to be approximately 4 in. from the footing edge, the center-to-center spacing will be

$$s = \frac{141 - 8}{11} = 12.1 \text{ in.}$$

This is an acceptable spacing in the 30 in. thick footing. If we drop to the No. 8 bars, the spacing becomes

$$s = \frac{141 - 8}{19} = 7.0 \text{ in.}$$

which is still not too crowded.

From these considerations it would be reasonable to reinforce the footing with any one of the three bar-size combinations from the No. 8 to the No. 10. Our preference would be for 16 No. 9 bars, which is one bar size below the limit for bond stress.

Actually, of course, this may not be the ideal footing design. If construction cost is the major determinant, the ideal footing would be the one with the lowest total costs for forming, concrete, and steel. Although our calculations have established the 30 in. dimension as the least possible thickness, it may be more economical to have more concrete and less reinforcing by using a thicker footing to increase the moment arm for the tension-steel calculation.

One possible limitation for the relationship between the concrete thickness and the steel area is the total percentage of steel in the bending cross section. If this is excessively low, the section is hardly being reinforced. Our recommendation is to bracket the steel percentage between one third and one half of 1%, or

$$p = (0.0033 \text{ to } 0.005)(bd)$$

For our footing this range is

$$A_s = (0.0033)(141 \times 25.9) = 12.1 \text{ in.}^2$$

to

$$A_s = (0.005)(141 \times 25.9) = 18.3 \text{ in.}^2$$

Our calculated requirement of 15.1 in.^{2} falls almost exactly in the middle of this range.

There are a number of other possible considerations that may affect the choice of the footing dimensions, such as the following:

1. ***Restricted height.*** Where the height (or footing thickness) is critical it may be desirable to use a pier or crossribs to substantially reduce the footing thickness. This limit may be because of

excavation problems, ground water conditions, the presence of compressible lower soil strata, or a major concern for footing weight when the allowable soil pressure is very low. Use of piers or pedestals is discussed in Section 3.5 and other means for reduction of the footing mass are discussed in Section 3.6.

2. ***Need for dowels.*** When the footing supports a reinforced concrete column, the dowels for the column reinforcing must be extended a required distance into the footing. This problem is discussed in Section 3.4.
3. ***Restricted footing width.*** Sometimes the proximity of other construction or close spacing of columns makes it impossible to use the required square footing. One solution is to use an oblong footing, the design of which is discussed in Section 3.7. When the restriction is severe and precludes the possibility of placing the footing symmetrically with the column, it may be necessary to use a strap, or cantilever footing, as discussed in Section 3.8.

3.4 Support Considerations

Objects that sit on footings may be attached to the footings in a number of ways. As shown in Figure 3.25, the basic types of attachment are as follows:

Direct bearing without anchorage. Direct bearing is the usual case with small, unreinforced masonry piers. No uplift resistance and little lateral load transfer are possible with this detail. Adhesion of mortar is the only actual attachment mechanism, although friction due to dead load will develop a lateral load transfer potential.

Doweling of reinforcing bars. Dowels provided for vertical reinforcement in concrete or masonry elements have the potential for providing uplift resistance and some lateral resistance.

Anchor bolts. Bolts are commonly used to attach elements of wood, metal, and precast concrete. In most cases the principal function of these bolts is simply to hold the supported elements in place during construction. However, as with dowels, there is a potential for development of resistance to uplift or lateral forces.

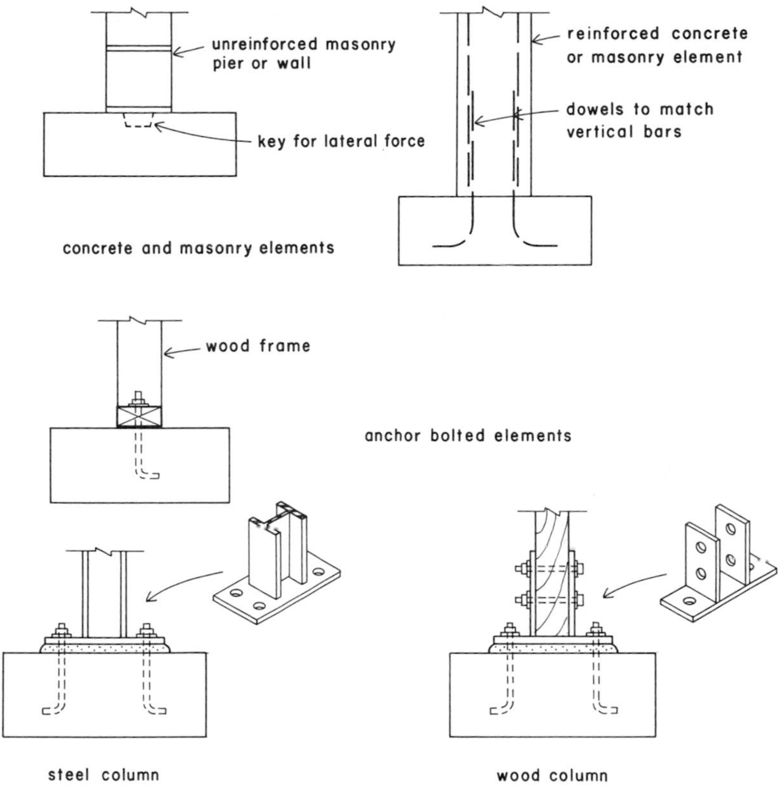

FIGURE 3.25. Common attachments to footings.

Special embedded anchors. Embedded anchors may be patented devices or custom-designed elements for various purposes. A variety of elements are available for attachment of wood columns. One special situation involves attaching elements to provide for subsequent removal, and possibly for repeated removal and reattachment. Another involves providing for future permanent attachment where addition to the structure is anticipated. Figure 3.26 shows a number of these types of attachment.

When a reinforced concrete column rests on a footing it is usually necessary to develop the vertical compressive stress in the column reinforcing

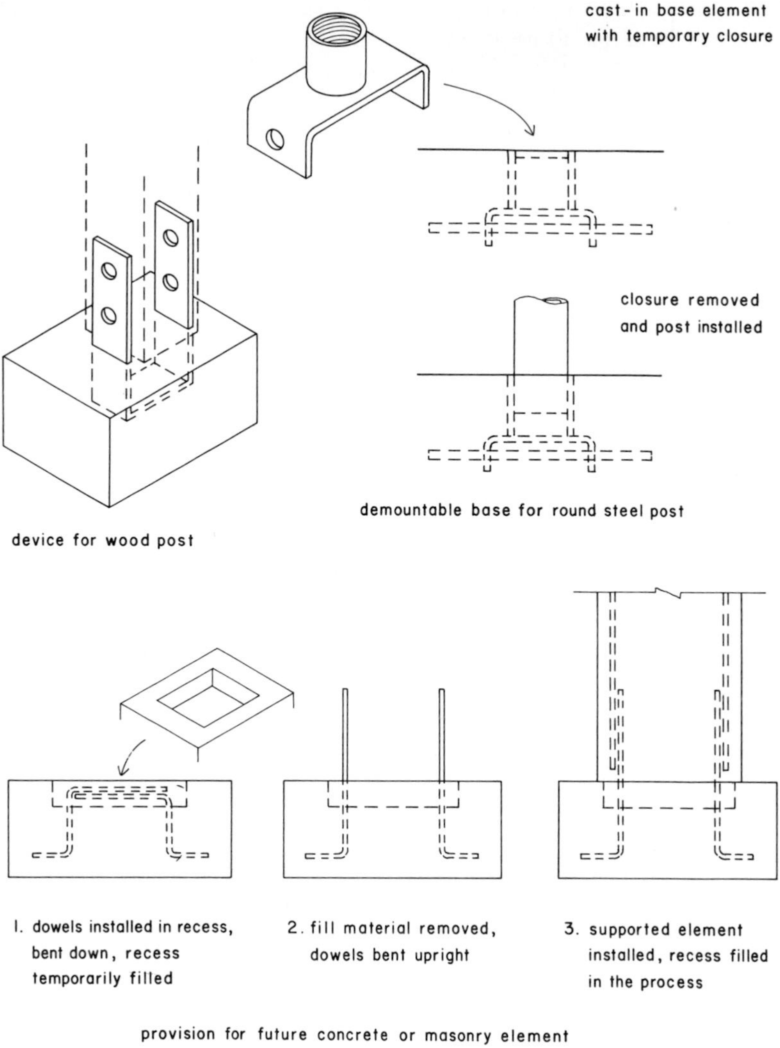

FIGURE 3.26. Special attachment devices for footings.

by doweling action into the supporting concrete structure. For reinforcing bars of large size and high yield strength, the distance required for this development length can become considerable and may be a major consideration for determining the footing thickness. The various requirements for this situation are given in Section 15.6 of the ACI code (Ref. 11).

Tension anchorage of reinforcing bars is commonly developed by a combination of dowel embedment action and hooking of the end of the bar. The ACI code also specifies the details of hooks.

A major construction detail problem with using anchor bolts and other embedded anchoring elements is the accuracy of their placement. Foundation construction in general is typically quite crude and not capable of high precision. It therefore becomes necessary to make some provision for this potential inaccuracy if elements to be attached later require relatively precise positioning. Leveling beds of grout are used to develop precise vertical positioning. Some of the techniques for providing horizontal adjustment are shown in Figure 3.27. It is possible to provide for adjustment of the anchor device itself, the object to be anchored, or a combination of both means.

It is also possible to attach elements to foundations after the foundation concrete is hardened through the use of drilled-in, driven, or explosively set anchoring devices. While not desirable for major structural elements, this is quite commonly done for items such as partitions, window and door frames, signs, and equipment bases. Figure 3.28 shows some of the types of devices used for these purposes.

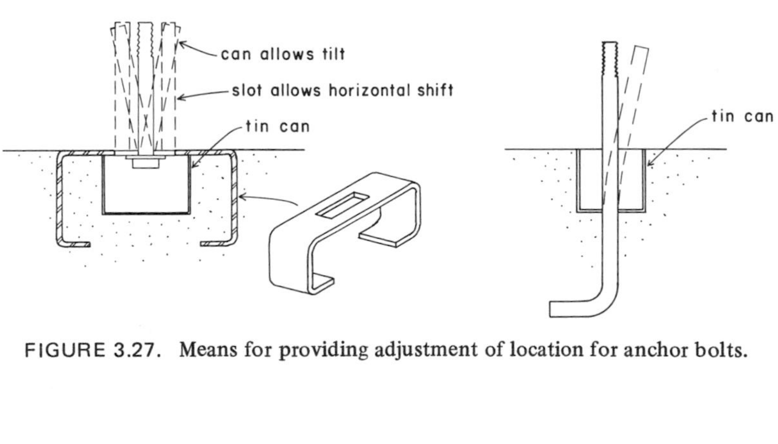

FIGURE 3.27. Means for providing adjustment of location for anchor bolts.

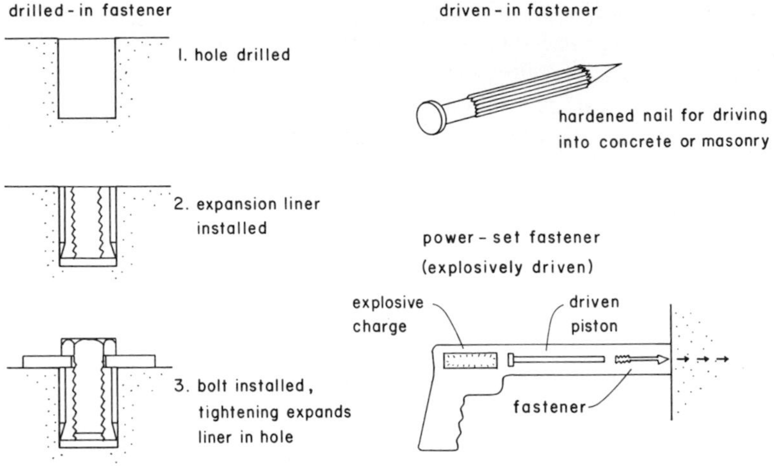

FIGURE 3.28. Attachment devices installed after casting of concrete.

3.5 Use of Piers

There are many instances in which a short compression element, called a pier or pedestal, is used as a transition between a footing and some supported element. As shown in Figure 3.29, some of the purposes for piers are the following:

To permit thinner footings. By widening the bearing area on the top of the footing the pier may achieve a reduction in shear and bending stresses in the footing, permitting the use of a thinner footing. This may result in cost savings as well as the reduction of total weight of the foundation.

To effect a transition from a column with high-strength concrete. In order to permit the use of a lower-strength concrete in the footing, it may be possible to use a pier with a strength that is transitional between the column and footing.

To achieve doweling for large, high-strength reinforcing bars. As discussed previously in regard to dowels, the use of a pier may allow the doweling of large column-reinforcing bars within the pier height. The pier reinforcing may then be done with smaller bars whose doweling requirements will be more modest for the footing to accommodate.

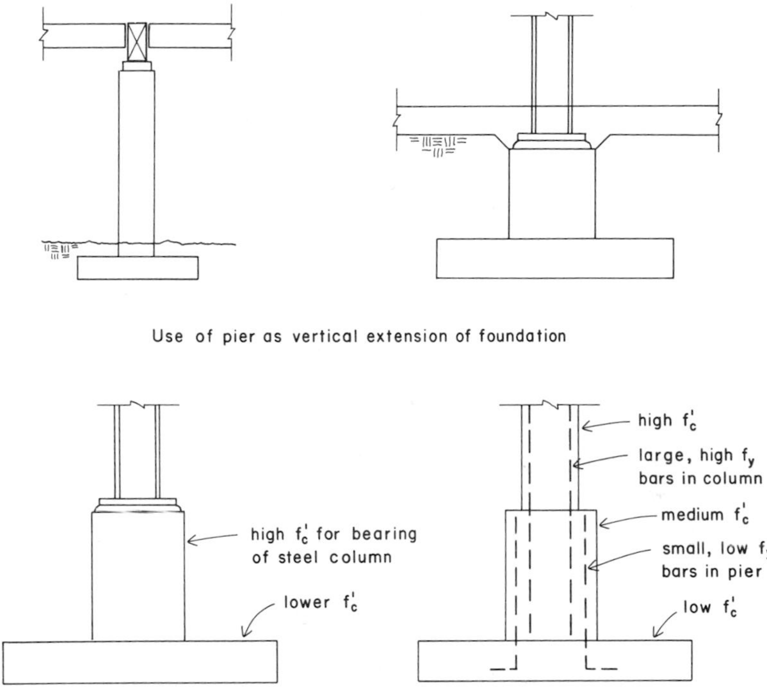

Use of pier as vertical extension of foundation

Use of pier to reduce footing thickness

FIGURE 3.29. Uses of piers with bearing footings.

This effect combined with the previous two effects can substantially reduce the structural requirements for the footing.

To keep wood and metal elements above the ground. When the bottom of the footing must be some distance below the ground surface, a pier may be used to keep elements above the ground if moisture and other effects are potentially damaging.

To provide support for elements at some distance above the footing. In addition to the previous situation, there are others in which the elements to be supported may be some distance above the footing. This may occur with tall crawl spaces, basements, or where the footing elevation must be dropped a considerable distance to obtain adequate bearing.

Piers of masonry or concrete are essentially short columns. When carrying major loads they should be designed as reinforced columns with appropriate vertical reinforcing, ties, and dowels. When loads are light and the pier height is less than three times the thickness, however, they may be designed as unreinforced piers. If built of hollow masonry units they should have the unit voids completely filled with lean concrete grout. Tables 3.7 and 3.8 give loads for short masonry and concrete piers. The table entries represent only a sampling of the potential sizes for these piers and are intended to give a general idea of the load range, not to be construed as standard or recommended sizes.

There are minimum as well as maximum heights for piers. If the pier is very short and the area of the bearing contact with the object supported by the pier is small, there will be considerable bending and shear in the pier, similar to that in a footing. This can generally be avoided if the pier is at least as tall as it is wide. For a pier of concrete it is theoretically possible to reinforce the pier in the same manner as a footing, although this is generally not feasible.

Masonry piers are subject to considerable variation. Primary concerns are for the type of masonry unit, the class of mortar, and the general type of masonry construction. For the piers in Table 3.7 we have assumed a widely used type of construction for piers when load magnitudes are relatively low. This involves the use of hollow concrete blocks with all voids filled with grout. For this construction we have used the allowable stresses in the *Uniform Building Code* (Ref. 9) as given in Table No. 24-B, assuming the use of Type S mortar. The allowable stress in compression on the gross cross sectional area of the pier is 150 psi. According to a footnote to the table, an increase of 50% is permitted for the stress at the bearing contact with the object supported by the pier. Based on these requirements it is possible to establish a maximum allowable load for a given size pier, and to derive the corresponding minimum column size on the basis of the ratio of the two allowable stresses. These two items are given in Table 3.7.

For supported objects with contact areas smaller than the minimum column size listed, the pier load will be limited by the contact area, rather than by the potential pier capacity. Thus the table includes pier capacities for a range of smaller columns, based on the stress limit for the contact area.

Although the piers in Table 3.7 are designed using criteria for unre-

TABLE 3.7. Unreinforced Masonry Piers

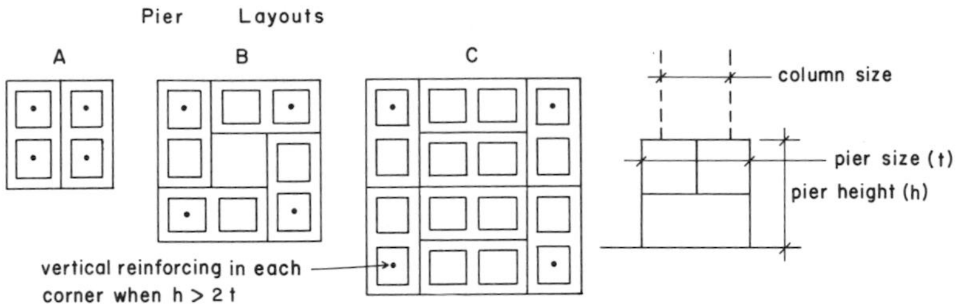

Nominal pier size (in.)		16	24	32
Pier layout (see illustration)		A	B	C
Maximum height = 3 t (in.)		48	72	96
Maximum allowable load (k) based on column size of:	8 in.	14.4	14.4	14.4
P = (column area) (0.225)	12 in.	32.4	32.4	32.4
	16 in.	—	57.8	57.8
	20 in.	—	—	90.0
	24 in.	—	—	129.6
Maximum allowable load (k) based on gross area of pier P = (pier area) (0.150)		38.4	86.4	153.6
Minimum column size required for development of maximum pier load (in.)		13	20	26
Recommended reinforcing		4 No. 3	4 No. 4	4 No. 5

inforced masonry, we recommend the use of vertical reinforcing in the four corners of the pier when the pier height exceeds twice the pier thickness. This reinforcing is quite minimal in cost and need not be doweled into the footing, but it adds a degree of toughness to the pier against the ravages of shrinkage, temperature stresses, and possible damage during construction of the supported structure. As stated pre-

TABLE 3.8. Unreinforced Concrete Piers

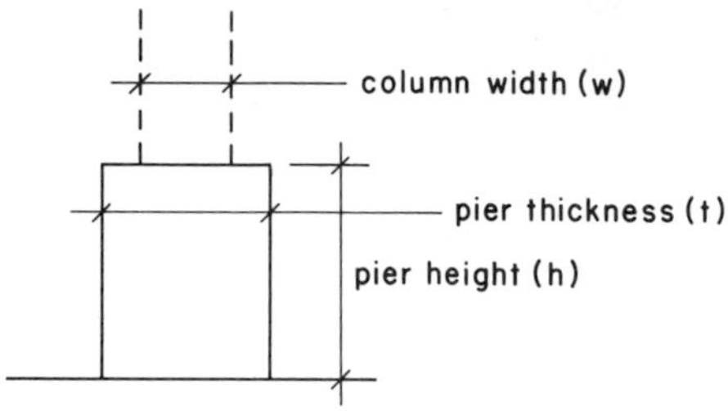

Pier Thickness *t* (in.)	Column Width *w* (in.)	Maximum Height 3 *t* (in.)	Allowable Load (k) for Concrete Strength (psi) as Shown 2000	3000	4000
10	8	30	40.6	60.1	81.2
12	8	36	45.3	68.0	90.6
12	10	36	61.4	92.1	122.8
16	8	48	48.0	72.0	96.0
16	10	48	73.6	110.4	147.2
16	12	48	95.6	143.4	191.2
16	14	48	115.2	172.8	230.5
20	10	60	75.0	112.5	150.0
20	12	60	106.5	159.8	213.1
20	14	60	135.6	203.4	271.2
20	16	60	162.8	244.2	325.6

viously, if the pier height exceeds three times its thickness, the pier should be designed as a reinforced masonry column with appropriate vertical reinforcing, horizontal ties, and footing dowels.

Concrete piers are also subject to considerable variation. As mentioned previously, if the pier height is less than the pier thickness and the size of the object is small with respect to the pier width, the pier may need to be investigated for footing-type action and reinforced appropriately. The entries in Table 3.8 assume the use of an unreinforced concrete pier. For this condition we recommend a minimum pier height equal to the pier width and a maximum height equal to three times the pier width.

For the unreinforced concrete pier there are also two stress considerations. The 1963 ACI code (Ref. 11—see the appendix) gives two allowable stresses for bearing on the pier. When the contact area with the supported object is one third or less of the pier cross sectional area, the allowable stress is 0.375 f'_c. When the contact area is equal to the full cross section of the pier, the stress is limited to 0.25 f'_c. For ratios of contact area to pier area between these values the allowable stress may be proportioned between these values. Figure 3.30 is a plot of these relationships and may be used to establish values for the allowable bearing stress for intermediate ratios of contact area to pier area.

In Table 3.8 the allowable loads for piers are based on the size of the supported element, using the relationships just discussed to establish the allowable bearing stress. Although theoretically possible, it is not

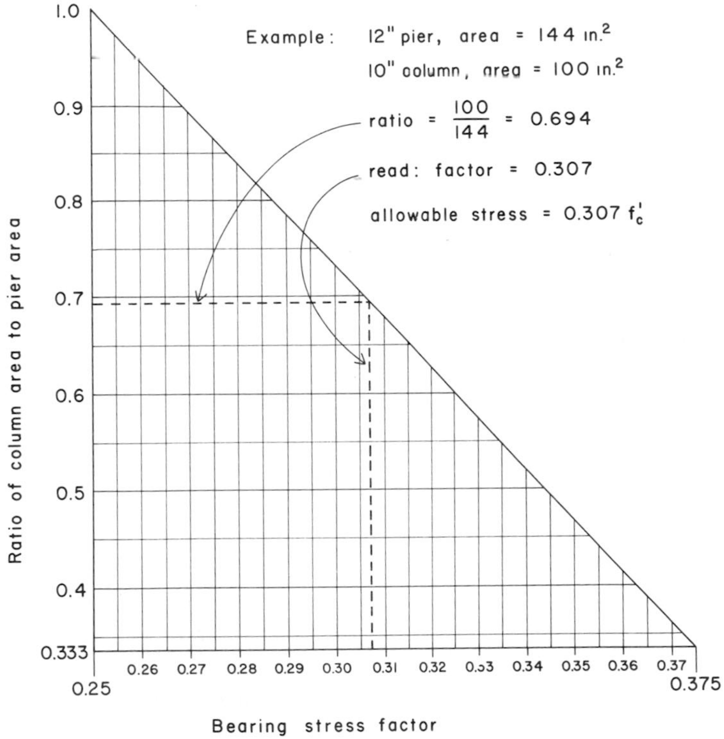

FIGURE 3.30. Allowable bearing stress for concrete piers, 1963 ACI Code.

likely that the contact area of the object supported by the pier will be the full area of the pier cross-section, since this would generally negate the concept of using the pier in the first place. In the table we have given the maximum pier capacities for an effective column width slightly less than the pier width. Bearing stress on the full pier cross section is more likely to be critical when the concrete strength of the footing is less than that of the pier. This condition may exist when the pier is used as a transition from a concrete column of very high strength concrete.

3.6 Reduction of Mass in Large Footings

As footings get very large it becomes advisable to utilize some means to reduce the concrete mass. The two basic techniques are using piers and using ribs to stiffen the footing. A large footing may support something other than a column load at its center, in which case the manner in which the supported elements are spread over the footing may reduce the effects of bending and shear in the footing. For the large column footing, however, the problem is simply a matter of reducing the cantilever effect or increasing the structural efficiency of the footing.

Figure 3.31 shows a number of possibilities for development of a bearing foundation for a concentrated load. At (*a*) is shown the use of a simple rectangular block, or pier, that is significantly larger than the supported element, so that the edge cantilever of the footing and the total perimeter of the support are both altered a substantial amount, permitting a major reduction in the footing thickness. Reduction in the footing reinforcing may not be as substantial, however, since the reduction in footing thickness will shorten the moment arm for the steel tension force.

At (*b*) is shown a modification of the pier into a semipyramidal form that permits a wider spread of the pier bottom without having the mass of the pier become excessive. The ultimate extension of this technique is shown at (*c*), where the pier and footing are merged into a single form. These forms are somewhat more difficult to construct than the simple rectangular blocks shown at (*a*), and their use and the details of their construction should be developed by someone with experience in the problems of forming and placing field-cast concrete.

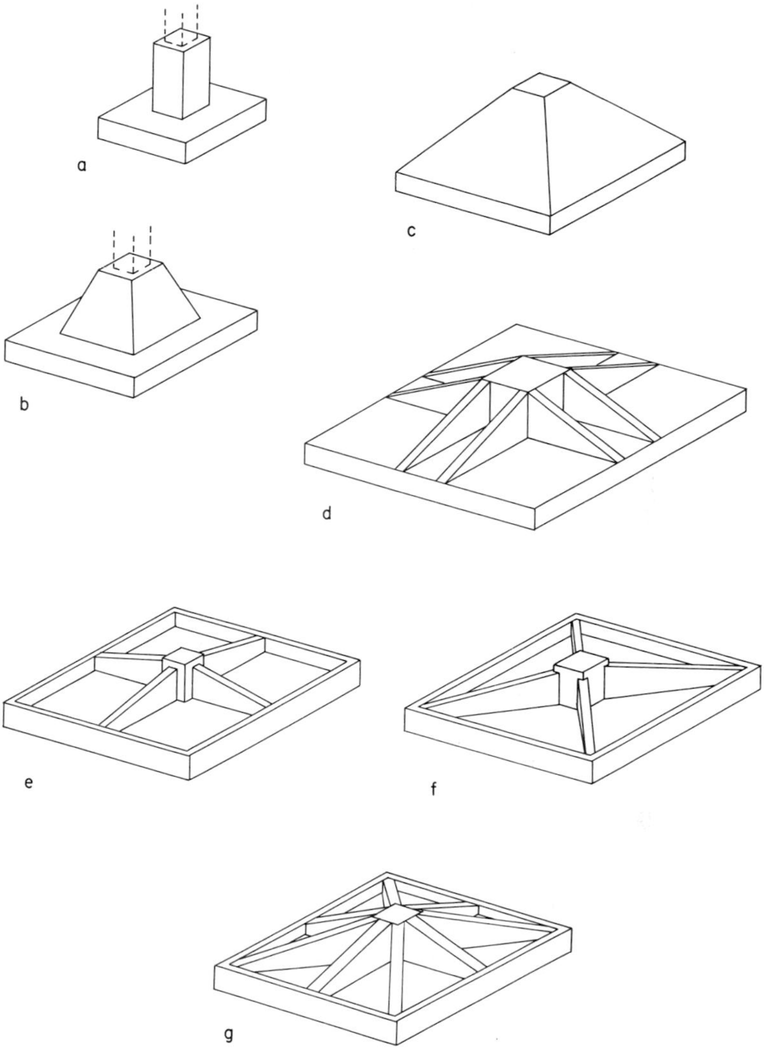

FIGURE 3.31. Techniques for reducing thickness and weight of column footings.

As was discussed previously, piers are also used where the distance between the bottom of the element requiring support and the level at which adequate bearing can be achieved is significantly greater than the required thickness of the footing. Therefore, if it is already established that a pier is to be used, and its required height is known, part of the geometry of the types of foundation shown in (*a*), (*b*), and (*c*) will be determined. The effects of the pier width and form can then be considered in conjunction with the resulting variations in the footing, and a total cost comparison of alternatives can be made including all the factors: forming, concrete, reinforcing, and special labor to place the concrete.

In special situations it may be desirable to reduce the footing weight by using some form of stiffening. The remaining illustrations in Figure 3.31 show various possibilities for this. At (*d*) and (*e*) are shown the use of ribs extended from a pier. At (*f*) and (*g*) stiffening ribs are added at the edge of the footing; this is a means for eliminating the cantilevering requirement for the footing. These techniques are generally feasible only when the footing mass must be reduced because of a very low allowable soil pressure. The ribs in such a footing will generally be heavily reinforced and will probably require a higher concrete strength than that ordinarily used for simple foundation construction. Considering the forming costs as well, it is not likely that this type of construction will save money.

When soil pressures recommended for design are as low as 2000 lb/ft^2 the use of a simple rectangular footing for large loads is questionable. Referring to Table 3.5, one notes that the largest entries for the lower soil pressures are quite small compared to those for the higher soil pressures. When the weight of the footing itself uses up a substantial percentage of the allowable soil pressure, the use of such a foundation becomes very questionable. One alternative is to utilize some of the means we have illustrated to reduce the footing mass. The other alternative is to consider the use of a deep foundation.

3.7 Rectangular Column Footings

When constraints on the size of a column footing prevent the use of the required square footing, it is sometimes possible to use a rectangular footing that is oblong. Although a square footing is actually also rectangular, the term "rectangular" is traditionally used to describe a footing with one plan dimension larger than the other. The close proximity of other elements of the building construction or property lines are the usual reasons for using such a footing.

An oblong footing must be designed separately for bending in two directions, called the long and short directions. The greater the difference between these two dimensions, the more the footing tends to act primarily in one-way bending. Regardless of the relative proportions, there tends to be a concentration of the bending in the short direction in the vicinity of the column. The ACI code provides for this latter phenomenon by requiring that the reinforcing found to be required for bending in the short direction must be proportioned to provide a specific percentage of the total within a zone equal to the width in the short direction. This procedure is demonstrated in the example that follows. Except for this requirement, the design of a rectangular footing generally follows the same procedures as those for a square footing.

Example 5 Rectangular Column Footing

We assume in this example that the column in Example 4 must be provided with a footing that is limited to a dimension of 9 ft in one direction. A preliminary investigation, either by calculation or reference to a table such as Table 3.5, would establish that the required square footing must be the 30 in. thick, 11 ft 9 in. square footing designed in Example 4. If the footing is to be limited to 9 ft in one direction it will therefore need to be considerably more than 11 ft 9 in. in the other direction. There is no particular limit on the proportions of the two dimensions, but the feasibility for such a footing generally limits the ratio of the two dimensions to approximately 2:1.

As given for Example 4, the data for the footing is as follows:

Column load: 500 k.

Column size: 15 in. square.

Maximum allowable soil pressure: 4000 lb/ft^2.

Concrete design strength: $f'_c = 3000$ psi.

Allowable tension on reinforcing: 20,000 psi.

The usual critical concrete stress limit for such a footing is that of beam-type shear in the long direction. In most cases, however, it is desirable to use a thickness sufficient to keep the amount of reinforcing in the long direction within a reasonable amount. This usually establishes a thickness greater than that limited by shear stress consideration. For a first design trial we will assume a thickness slightly larger than that required for the square footing.

Try: $h = 36$ in.

Then: footing weight $= \frac{36}{12}(150) = 450$ lb/ft^2

net usable soil pressure $= 4000 - 450 = 3550$ lb/ft^2

$$\text{required area} = \frac{500{,}000}{3550} = 140.8 \text{ ft}^2$$

$$\text{required long dimension} = \frac{140.8}{9} = 15.64, \text{ say 15 ft 8 in.}$$

actual area $= (9)(15.67) = 141.0$ ft^2

$$\text{design soil pressure} = \frac{500{,}000}{141.0} = 3546 \text{ lb/ft}^2$$

Although the square footing was designed for an average d, halfway between the actual d in the two directions, the reinforcing for the rectangular footing will be designed for the true d in the two directions. Since the reinforcing in the long direction will usually be considerably greater, it will be given the advantage of being placed in the lower layer, giving it the larger d. Assuming a No. 10 bar size, the d for the long direction will thus be

$$d = 36 - 3 - \frac{1.270}{2} = 36 - 3.635 = 32.365, \text{ say 32.4 in.}$$

Analysis for beam-type shear in the long direction is as follows. As shown in Figure 3.32, the section for this stress will be at the point 54.1 in. from the end of the footing. The total shear force on this section is thus

$$V = (3{,}546)(9)\left(\frac{54.1}{12}\right) = 143{,}879 \text{ lb}$$

and the shear stress is

$$v = \frac{146{,}879}{(108)(32.4)} = 42 \text{ psi}$$

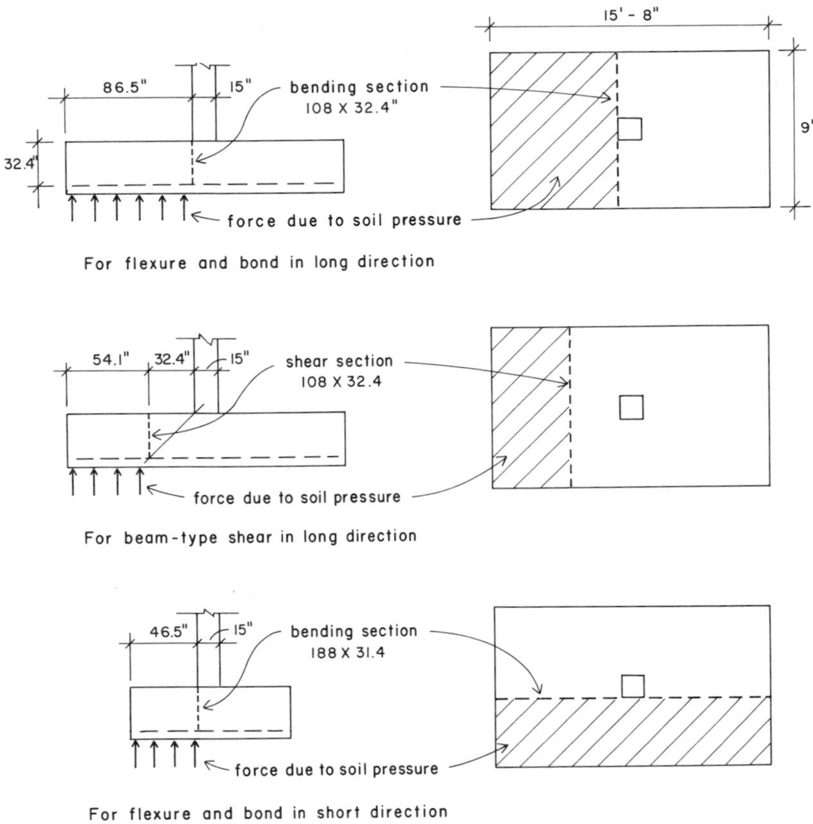

FIGURE 3.32. Assumptions for stress analysis in the rectangular column footing–Example 5.

which is compared to the allowable stress of

$$v = 1.1\sqrt{f_c'} = 1.1\sqrt{3000} = 60 \text{ psi}$$

Comparison with the investigation in Example 4 should show that the peripheral shear will not be critical with the increased footing thickness. However, if any doubt exists, the investigation should be made in the same manner as in Example 4. Since the beam-type shear is lower than the limit, it is possible to reduce the footing thickness slightly. However, we will stay with the 36 in. thickness for reasons discussed previously.

For bending and bond stress considerations the critical section will be at the face of the column and the cantilever distance will be 86.5 in., as shown in Figure 3.32. The vertical force and bending are thus determined as

$$V = (3{,}546)(9)\left(\frac{86.5}{12}\right) = 230{,}047 \text{ lb}$$

$$M = (230{,}047)\left(\frac{86.5}{2}\right) = 9{,}949{,}532 \text{ lb-in.}$$

For comparison, the balanced stress moment capacity of the section is

$$M_R = Rbd^2 = (226)(108)(32.4)^2 = 25{,}622{,}542 \text{ lb-in.}$$

Since this is approximately three times the required moment, we may assume the concrete bending stress to be quite low and may use a conservative approximation for the true j value. The amount of tension reinforcing is thus determined as

$$A_s = \frac{M}{f_s jd} = \frac{9{,}949{,}532}{(20{,}000)(0.9)(32.4)} = 17.06 \text{ in.}^2$$

A possible selection would be 14 No. 10 bars with a total area of

$$A_s = (14)(1.27) = 17.78 \text{ in.}^2$$

We must also consider the bond stress on the bars and the problems of bar spacing, as discussed in previous examples. Table 3.9 gives a num-

TABLE 3.9. Possible Bar Combinations—Long Direction (Example 5)

Bars (Number and Size)	Actual A_s Provided (in.^{2})	Total Perimeter Σ_0 (in.)	Actual Bond Stress $u = \frac{V}{\Sigma_0 jd}$ (psi)	Allowable Bond Stress $u = \frac{4.8\sqrt{f_c'}}{D}$ (psi)	Approximate Center-to-Center Spacing (in.)
22 No. 8	17.38	69.12	114	263	4.76
18 No. 9	18.00	63.79	124	233	5.88
14 No. 10	17.78	55.86	141	207	7.69
11 No. 11	17.16	48.73	162	186	10.00

ber of bar combinations that could be used to satisfy the area requirement for tension stress, together with the bond stresses and actual bar spacings. While it should not be considered an ideal choice, we would probably select 14 No. 10 bars.

As shown in Figure 3.32, the cantilever distance for bending in the short direction is 46.5 in., and the bending force and moment are as follows:

$$V = (3{,}546)(15.67)\left(\frac{46.5}{12}\right) = 215{,}318 \text{ lb}$$

$$M = (215{,}318)\left(\frac{46.5}{2}\right) = 5{,}006{,}143 \text{ lb-in.}$$

With the No. 10 bars in the long direction, and assuming the use of No. 6 bars in the short direction, the effective d is thus

$$d = 36 - 3 - 1.27 - \frac{0.75}{2} = 36 - 4.645 = 31.355, \text{ say } 31.4 \text{ in.}$$

and the area of tension reinforcing required is

$$A_s = \frac{M}{f_s jd} = \frac{5{,}006{,}143}{(20{,}000)(0.9)(31.4)} = 8.86 \text{ in.}^2$$

If No. 6 bars are used, this could be satisfied with 20 No. 6 bars, providing a total of

$$A_s = (20)(0.44) = 8.80 \text{ in.}^2$$

and resulting in a bond stress of

$$u = \frac{V}{\Sigma_0 jd} = \frac{215{,}318}{(20 \times 2.356)(0.9)(31.4)} = 162 \text{ psi}$$

which is compared to the allowable bond stress of

$$u = \frac{4.8\sqrt{f_c'}}{D} = \frac{4.8\sqrt{3000}}{0.75} = 350 \text{ psi}$$

The ACI code requires that this reinforcement be concentrated in the vicinity of the column by specifying that a certain percentage of the total reinforcement required shall be placed in a width equal to the

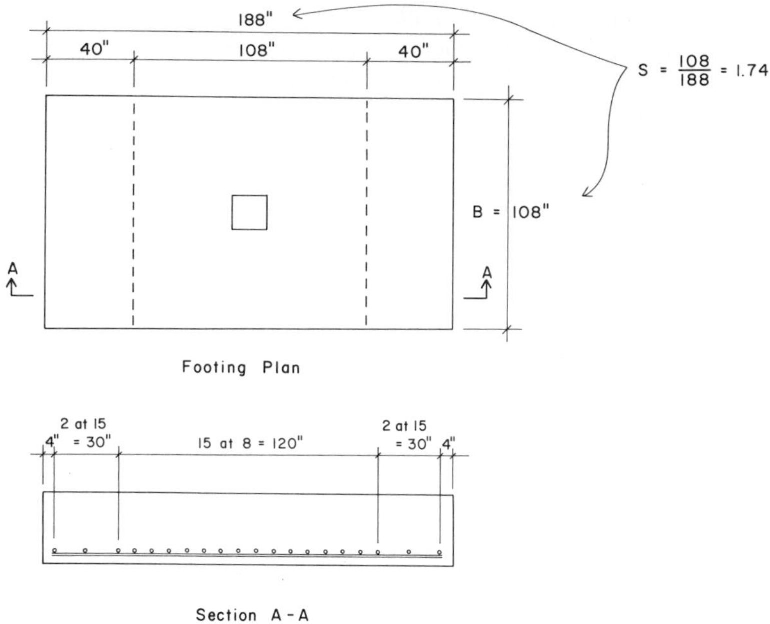

FIGURE 3.33. Development of the reinforcing in the short direction—Example 5.

short dimension of the footing, as shown in Figure 3.33. The required percentage is established by the following:

$$\frac{\text{reinforcement in width } B}{\text{total reinforcement}} = \frac{2}{S+1}$$

in which S is the ratio of the long dimension to the short dimension for the footing. Thus for our example

$$S = \frac{15.67}{9} = 1.74$$

and using the total of 20 bars

$$\text{number of bars required in } B = (20)\left(\frac{2}{2.74}\right) = 14.6, \text{ or } 15 \text{ bars}$$

A possible spacing of the bars in the short direction, giving reasonable compliance with the code requirement, is shown in Figure 3.33.

Comparison with the square footing in Example 4 shows that this footing uses approximately 23% more concrete but almost the same total weight of steel bars. If a slightly thinner footing were used, as would be possible within the concrete stress limits, the amount of additional concrete would be less, but the amount of reinforcing would be increased.

3.8 Combined Column Footings

When two or more columns are quite close together, it is sometimes desirable or necessary to use a single footing, as shown in Figure 3.34. If the columns and their loads are symmetrical, the footing is usually designed to function in its long direction as a uniformly loaded double cantilever beam, as shown in the illustration. If the column loads are not equal, the usual procedure is to find the location of the centroid of the column loads and to design a footing whose plan centroid coincides with that of the column loads. This will assure a uniform soil pressure on the bottom of the footing. There are many possible variations for such a combined footing. The following example illustrates the design of a simple rectangular footing for a symmetrical loading.

EXAMPLE 6 COMBINED FOOTING WITH TWO EQUAL LOADS

Design data and criteria:

Column loads: 300 k each.
Column spacing: 8 ft center-to-center.
Maximum allowable soil pressure: 4000 lb/ft^2.
Concrete design strength: $f_c' = 3000$ psi.
Allowable tension on reinforcing: 20,000 psi.

Table 3.5 shows that the required square footing will be approximately 9 ft on a side. Since this is not possible in this case, the chief options are either to use two rectangular footings, or to use a combined footing.

Opting for the combined footing, our initial decisions are for the footing thickness and plan dimensions. There is no single optimal combination for these dimensions, although other factors in the foundation design or the details of the building construction may influence the

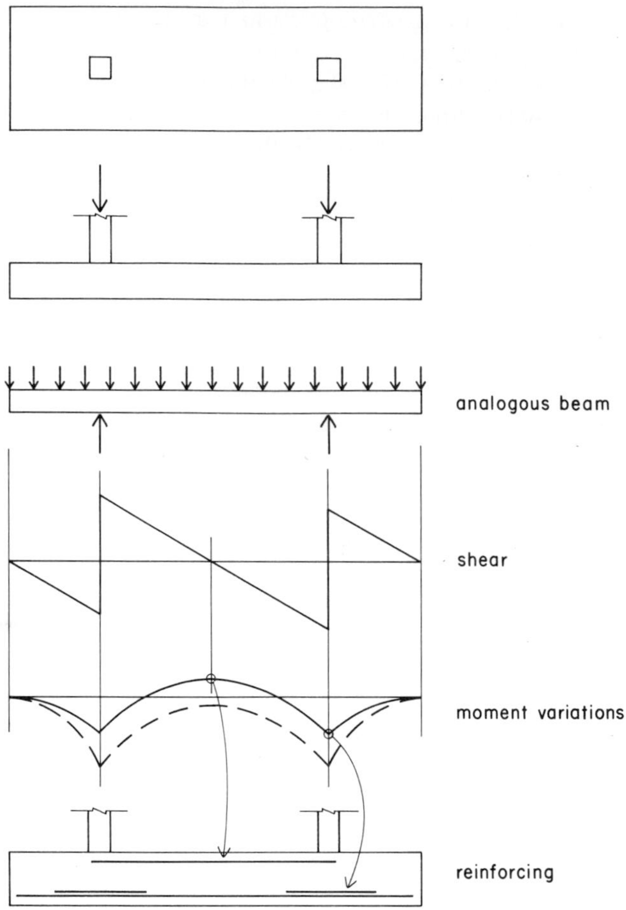

FIGURE 3.34. Actions of a symmetrical combined footing for two columns.

choices. We will arbitrarily assume that a thickness approximately equal to that for a square footing is a reasonable first try.

Try: $h = 24$ in.

Then: footing weight $= \frac{24}{12}(150) = 300$ lb/ft^2

net usable soil pressure $= 4000 - 300 = 3700$ lb/ft^2

$$\text{required footing area} = \frac{(300{,}000)(2)}{3700} = 162.2 \text{ ft}^2$$

If the footing were square, this would require a side dimension of

$$w = \sqrt{162.2} = 12.73 \text{ ft}$$

For various reasons this is probably not the desirable footing plan. In this case the cantilever distance from the column centers to one edge would be over 6 ft, while in the other direction it would be only slightly over 2 ft. Bringing these two cantilever distances reasonably close is one possible design guide. We may thus consider other possible plan dimension combinations. For an oblong footing, referring to the short dimension as w and the long dimension as l, some possible combinations are the following:

$$l = 14, w = \frac{162.2}{14} = 11.59$$
$$l = 15, w = 10.81$$
$$l = 16, w = 10.13$$
$$l = 17, w = 9.54$$
$$l = 18, w = 9.01$$
$$l = 19, w = 8.54$$
$$l = 20, w = 8.11$$

We will choose the combination of an 18 ft length and a 9 ft width. With no other dimensional constraints known, this gives approximately equal cantilevers and is a situation that will result in a footing similar to an equivalent square one, with reinforcing approximately the same in both directions. With the chosen size the actual net design soil pressure will be

$$p = \frac{(300{,}000)(2)}{18 \times 9} = 3704 \text{ lb/ft}^2$$

For bending and shear in beam-type action in the long direction the footing is designed for a uniform load of

$$w = (9)(3704) = 33{,}336 \text{ lb/ft}$$

For simplification we will consider a load of 33.4 k/ft. With this loading the 18 ft double-cantilever beam will have the shear and moment diagrams as shown in Figure 3.35. Before proceeding with the design of reinforcing, we must verify that the footing thickness chosen is adequate for the flexural and shear stresses in the concrete. Analysis

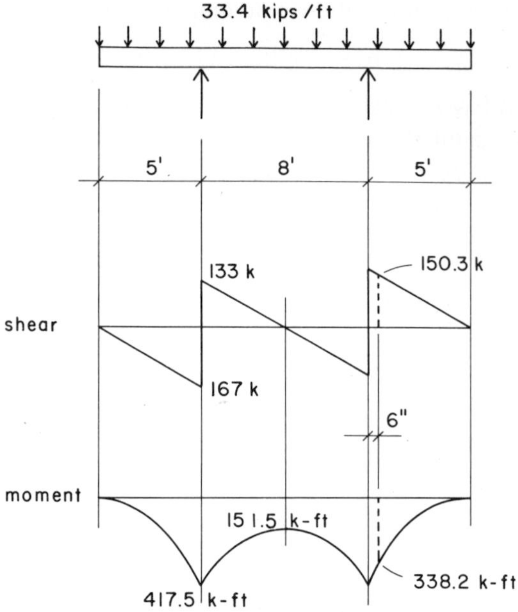

FIGURE 3.35. Beam-action loading for the combined footing–Example 6.

for this is essentially the same as for the square column footing, with investigations of beam-type and peripheral shear and a comparison of the calculated moment with the balanced stress moment capacity of the footing section. In our case the footing plan creates a situation very close to that for an equivalent 10 ft square footing. Referring to Table 3.5, we may therefore conclude that the thickness is not critical for the concrete stresses, as long as our column size is not smaller than that listed in the table.

We will choose to place the reinforcing in the long direction in the bottom layer, since the cantilever is slightly larger in this direction. We then assume an approximate effective depth to be h - 3.6 in., or 24 - 3.6 = 20.4 in. If we assume the column size to be 12 in., we may use the shear and moment values at a distance of 6 in. from the column centerline. As shown on the graphs in Figure 3.35, these design values are

V = 150.3 k, for bond stress

M = 338.2 k-ft, for flexural stress and reinforcing area

The area of steel required for the maximum moment condition is thus

$$A_s = \frac{M}{f_s jd} = \frac{(338.2)(12)}{(20)(0.9)(20.4)} = 11.05 \text{ in.}^2$$

This may be satisfied by using 14 No. 8 bars with a total area of 11.06 in.^{2}. The bond stress on these bars will be as follows:

$$u = \frac{V}{\Sigma_0 jd} = \frac{150{,}300}{(14 \times 3.142)(0.9)(20.4)} = 186 \text{ psi}$$

which is less than the allowable stress of 263 psi (see appendix, Table A.2).

In the 9 ft wide footing, the 14 No. 8 bars will be spaced at approximately

$$s = \frac{108 - 8}{13} = 7.69 \text{ in. center-to-center}$$

In the short direction the cantilever distance from the face of the column is only slightly shorter than that at the ends in the long direction (4 ft in the short direction versus 4 ft 6 in. in the long direction, assuming a 12 in. square column). Thus the bending and bond requirements will be close to those required in the long direction.

The particular form of the shear and moment graphs for this footing, shown in Figure 3.35, are a result of the column loads, column spacing, soil pressure, and our selection of the footing plan dimensions. As a result of all of these, the moment at the midpoint between the columns remains negative, and thus the location of the required reinforcing at this point is still in the bottom of the footing. If the columns had been farther apart, or our selection of the footing plan dimensions been for a shorter and wider footing, the reduced end cantilevers may have resulted in a positive moment value at midspan. In that case the required reinforcing at midspan would have been located in the top of the footing. Although it is not always possible to avoid, this situation is less desirable in the interests of simplifying the construction work.

When the closely spaced column loads are not equal and it is desired to maintain a uniform soil pressure on the footing, the footing plan form or location must be manipulated so that the centroid of the loads coincides with the centroid of the footing plan area. The following example illustrates this situation.

Example 7 Combined Footing With Unequal Column Loads

Design data and criteria are the same as Example 6, except:

Column load 1: 350 k.
Column load 2: 250 k.

The effect of this modification, in comparison with Example 6, is to shift the centroid of the loads slightly away from the midpoint between the columns and toward the higher-loaded column. However, the total load on the footing remains the same, and if the same footing thickness is used, the required area for the footing will be the same. We will assume, therefore, that it is still reasonable to consider using the 18 ft by 9 ft footing, if we simply shift its position so that its center coincides with the new load centroid.

As shown in Figure 3.36, this results in unequal end cantilevers of 4.33 and 5.67 ft. With the same 12 in. column, the design moments and shears for the two cantilevers are thus:

For the 4.33 ft end: $V = 127.9$ k, $M = 245.0$ k-ft.
For the 5.67 ft end: $V = 172.7$ k, $M = 446.4$ k-ft.

The moment values and corresponding steel area requirements at the two ends are thus 27% less and 32% greater respectively than in the symmetrical footing in Example 6. If the 24 in. footing thickness is retained, this means that we need about one third more reinforcing at the long end and one fourth less reinforcing at the short end. Or, with reference to the 14 No. 8 bars selected for the symmetrical footing, we could use

At the long end: $(1.32)(14) = 18.48$, or 19.
At the short end: $(0.73)(14) = 10.22$, or 11.

and place them as shown in Figure 3.37.

Peripheral shear stress in the concrete will be the same in this example, since the total column periphery, the total column load, the footing area, and the soil pressure are the same as in Example 6. However, the beam-type shear at the long end will be greater and should be inves-

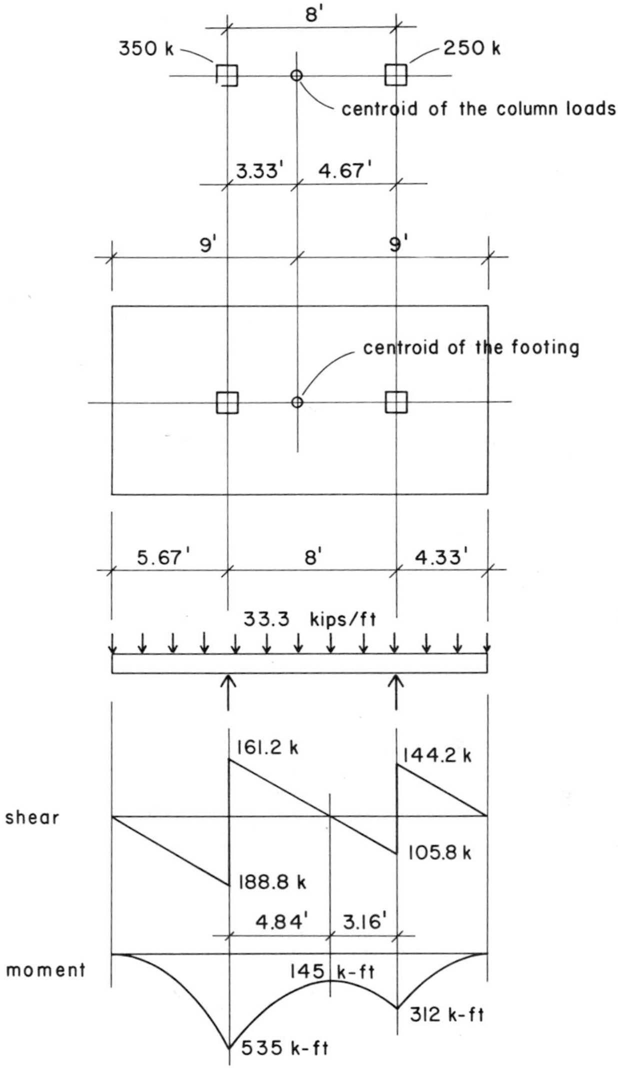

FIGURE 3.36. Development of centroids and beam actions for the combined footing–Example 7.

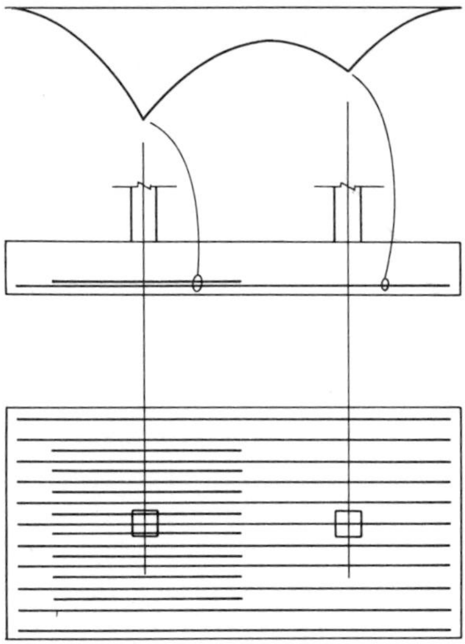

FIGURE 3.37. Possible placement of reinforcing–Example 7.

tigated. The critical section for this stress will be at a distance d from the face of the column, or

$$5.17 - \frac{20.4}{12} = 5.17 - 1.70 = 3.47 \text{ ft}$$

from the end of the footing.

With the loading of 33.3 k/ft the shear force at the section will thus be

$$(3.47)(33.3) = 115.6 \text{ k}$$

and the shear stress will be

$$v = \frac{115{,}600}{(108)(20.4)} = 52.5 \text{ psi}$$

which is less than the allowable stress of 60 psi for the concrete with f'_c of 3000 psi.

For the unequal column loads it is also possible to consider some alternatives to the off-center rectangular footing. Figure 3.38 shows

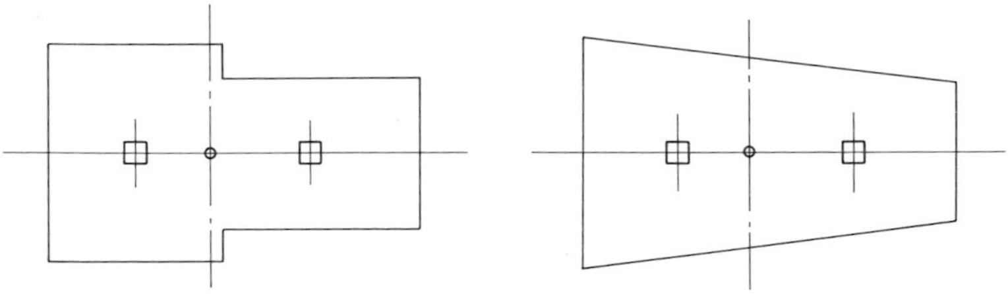

FIGURE 3.38. Plan variations for the unsymmetrical combined footing.

two variations that consist of manipulation of the plan form of the footing. In the first illustration a T-shaped plan is developed by first determining the typical square footing for the larger column load and then adding a symmetrical rectangular footing under the smaller column. This will automatically result in a T whose centroid coincides with that of the column loads.

In the second illustration a trapezoidal shape is developed so that the centroid of the trapezoid coincides with the centroid of the column loads. As with the simple rectangular footing, there are many possible variations of this form.

These variations—or other possible ones—are most likely feasible only where there are some constraining conditions that rule against the use of a simple rectangular form. The added costs for more complex forming and reinforcing will usually nullify any gains in reduced concrete volume or steel weight.

Occasionally, because of property line location, adjacent construction, or excavation difficulties, it is not possible to allow a footing to extend beyond a column on one side. A possible solution to this problem is to use a so-called *strap*, or *cantilever* footing. This technique consists of developing a combined footing for the restricted column and an adjacent column. This situation usually occurs because of columns at the building edge with restrictions because of the property line or an adjacent building.

Figure 3.39 shows the two most common forms for such a footing shared by an edge column and an interior column. In the upper illustration a simple rectangular footing is provided with its centroid developed to coincide with that of the total load on the footing. In the lower illus-

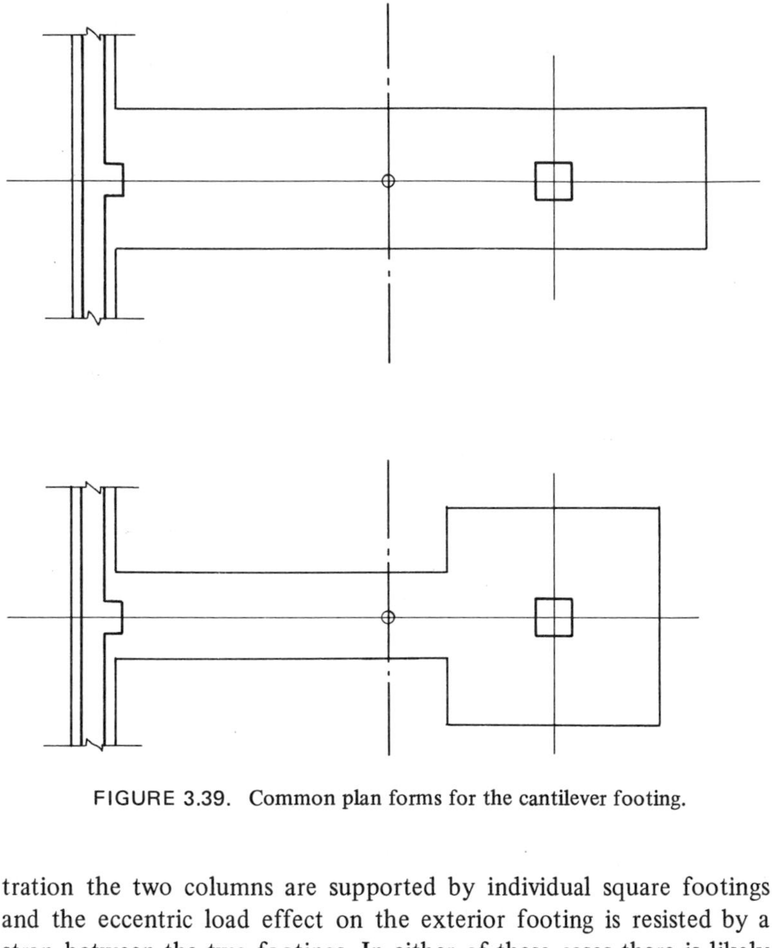

FIGURE 3.39. Common plan forms for the cantilever footing.

tration the two columns are supported by individual square footings and the eccentric load effect on the exterior footing is resisted by a strap between the two footings. In either of these cases there is likely to be considerable bending and shear in the footing between the two columns unless the columns are very closely spaced. If the distance between the two columns is great, it is usually necessary to use a construction such as that shown in Figure 3.40, in which a T-form is developed for the span between the two columns. The stem portion of this T would be heavily reinforced for the shear and bending occurring in the span.

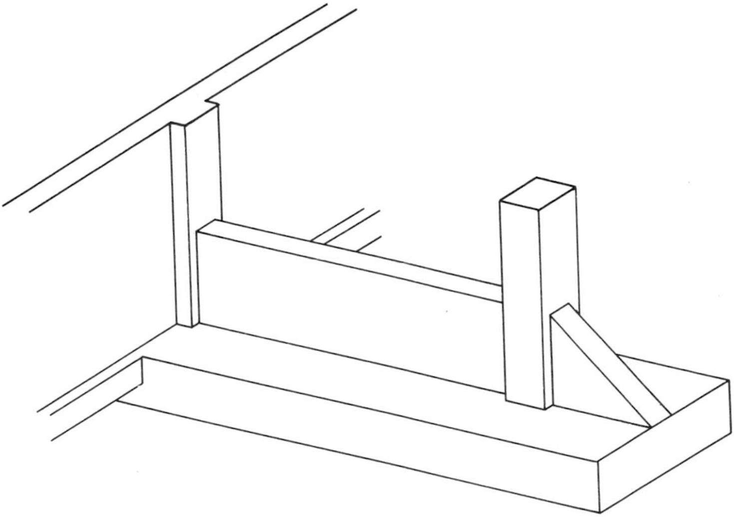

FIGURE 3.40. Use of a stiffening element for the cantilever footing.

EXAMPLE 8 CANTILEVER FOOTING

Design data and criteria:

Column loads and location as shown in Figure 3.41.
Maximum allowable soil pressure: 4000 lb/ft^2.
Concrete design strength: $f'_c = 3000$ psi.
Allowable tension on reinforcing: 20,000 psi.

If a simple rectangular footing is desired, as shown in Figure 3.41, the long dimension is established by the centroid of the loads. In the example this centroid is located 13.83 ft from the outside of the building, and the long dimension of the footing is obtained by doubling this dimension to 27.67 ft, or 27 ft 8 in. The other dimension for the footing plan can then be established once the value for the net usable soil pressure is established.

The required footing thickness in this case must be a raw guess for a first try.

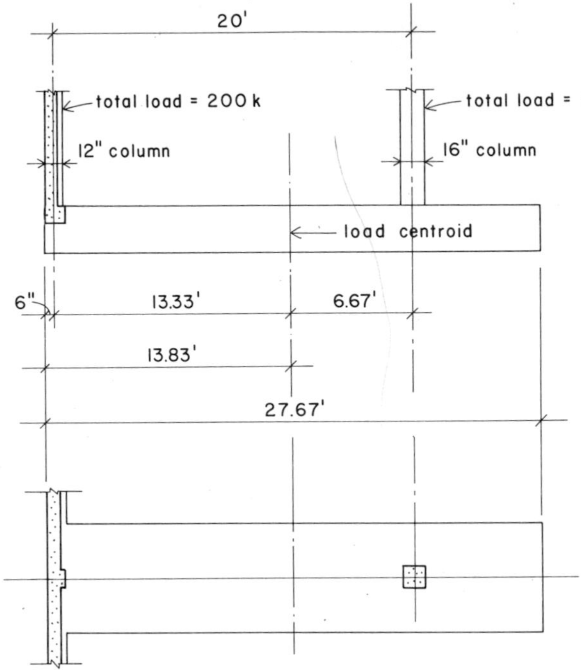

FIGURE 3.41. Design conditions and development of centroids for the cantilever footing–Example 8.

Try: $h = 36$ in.

Then: footing weight $= \frac{36}{12}(150) = 450$ lb/ft²

net usable soil pressure $= 4000 - 450 = 3550$ lb/ft²

$$\text{required footing area} = \frac{600{,}000}{3550} = 169.0 \text{ ft}^2$$

$$\text{required width} = \frac{169.0}{27.67} = 6.11 \text{ ft, say 6 ft 2 in.}$$

$$\text{design soil pressure} = \frac{600{,}000}{27.67 \times 6.17} = 3514 \text{ lb/ft}^2$$

In the long direction of the footing the column loads produce a linear loading of

$$w = \frac{600{,}000}{27.67} = 21{,}684 \text{ lb/ft}$$

Rounding this load off to 21.7 k/ft, the resulting shear and moment diagrams for the footing are as shown in Figure 3.42. The maximum beam-type shear stress occurs on the left side of the interior column. We have assumed an approximate d of 32 in. in locating this point for the critical shear stress section. With No. 11 bars and the long direction reinforcing placed on the bottom layer, the actual d will be closer to 32.3, which we will use for the stress calculations. Thus the maximum design shear stress is

$$v_c = \frac{172{,}500}{74 \times 32.3} = 72.2 \text{ psi}$$

which is in excess of the allowable stress of

$$v_c = 1.1\sqrt{f_c'} = 1.1\sqrt{3000} = 60 \text{ psi}$$

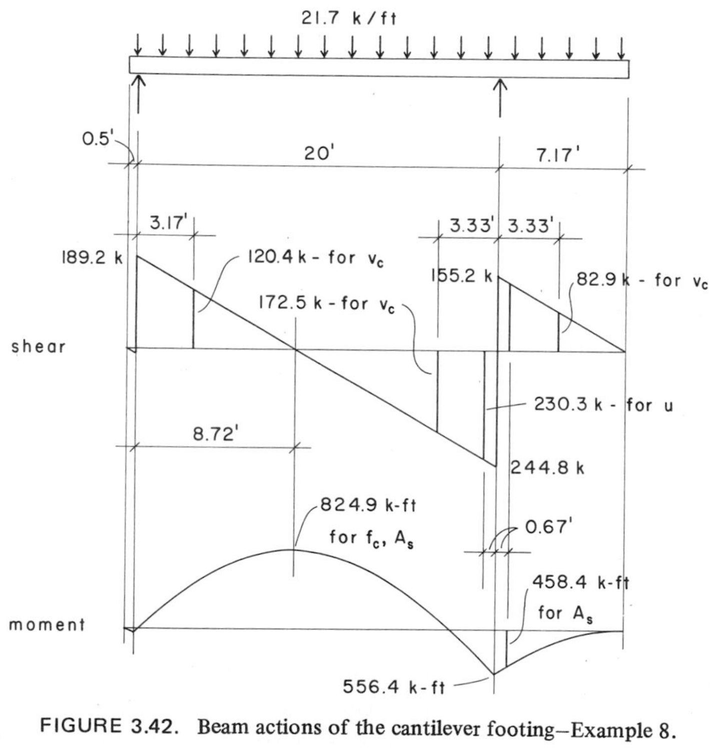

FIGURE 3.42. Beam actions of the cantilever footing–Example 8.

Thus, if the 36 in. thickness is retained, it will be necessary to provide shear reinforcing in this portion of the footing. The shear stresses at the other two sections may be proportioned as follows:

$$V = 116.9, v_c = \frac{120.4}{172.5}(72.2) = 50.4 \text{ psi}$$

$$V = 82.9, v_c = \frac{82.9}{172.5}(72.2) = 34.7 \text{ psi}$$

which indicates that shear reinforcing is not required at these locations.

The options at this point are to increase the footing thickness by an amount sufficient to reduce the concrete shear stress below 60 psi, or to use some vertical reinforcing to take the excess shear. While shear reinforcing is not generally used in footings, the very deep and narrow footing cross section is quite beamlike in its proportions, and it is reasonable to consider the use of some form of shear reinforcing. It is also possible that the use of some vertical reinforcing in the spanning portion of the footing may be the most practical means for supporting the reinforcing required in the top of the footing at this location.

The other major concern regarding the adequacy of the footing thickness is the very large bending moment of 824.9 k-ft. This may be compared to the balanced stress moment capacity of the cross section, which is

$$M_R = Rbd^2 = \frac{(0.226)(74)(33.3)^2}{12} = 1545 \text{ k-ft}$$

indicating that the flexural stress in the concrete is not critical.

Note that we have used a slightly larger value for d in this calculation, because the cover on the reinforcing is usually 1 in. less on the top and sides of the footing. The area of steel required for this moment is

$$A_s = \frac{M}{f_s jd} = \frac{(824.9)(12)}{(20)(0.9)(33.3)} = 16.51 \text{ in.}^2$$

Table 3.10 gives a number of bar size combinations that can be used to provide this total area, together with the approximate bar spacing, assuming the first outside bar to be approximately 4 in. from the edge. Any of the combinations in the table are probably all right. The spacing of the smaller bars is a bit tight, especially for reinforcing in the top of the footing, but it may be possible to leave a few of these bars out

TABLE 3.10. Possible Bar Combinations (Example 8)

Bars (Number and Size)	Actual A_s Provided (in.^{2})	Approximate Center-to-Center Spacing (in.)
21 No. 8	16.59	3.3
17 No. 9	17.00	4.1
13 No. 10	16.51	5.5
11 No. 11	17.16	6.6

while pouring the concrete and put them in place when the concrete is placed up to a level just below the top of the footing.

For the lower moment at the cantilever end of the footing the reinforcing area required is

$$A_s = \frac{(458.4)(12)}{(20)(0.9)(32.3)} = 9.46 \text{ in.}^2$$

Since the reinforcing is in the bottom of the footing, the bar spacing is somewhat critical. If these bars are checked for the critical shear force of 230.3 k, the bond stress will be quite high, so it is probably advisable to use a relatively small bar size. If 12 No. 8 bars are used, the total area provided will be 9.48 in.^{2}, the bond stress will be 210 psi, and the approximate center-to-center spacing will be 6 in.

In the short direction the cantilever distance is so short and the depth of the footing so great that a very small area of steel is actually required. If the footing cross section is reinforced as shown in Figure 3.43, the

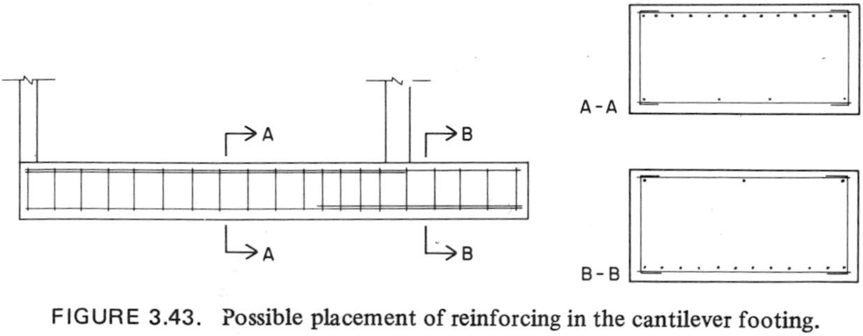

FIGURE 3.43. Possible placement of reinforcing in the cantilever footing.

bottom crossbars shown may be sized for the crossbending, although their real, practical purpose is to assist in placing and support of the long bars and the vertical bars.

There are many possibilities for variation of the combined footing discussed in the preceding example. The footing plan may be other than a simple rectangle; the T and trapezoidal shapes are two alternatives, as was discussed for the combined footing for the closely spaced columns (Example 6). It may also be possible to use a vertical stem in combination with the flat footing to produce a spanning element with a T-shaped vertical cross section, as illustrated in Figure 3.38.

The principal reason for considering these, or other, alternatives to the simple rectangular footing would be to increase the feasibility or reduce the cost of the footing. There may, however, be other motivations stemming from details of the construction or architectural planning. If a wall is to be placed between the two columns, it may be possible to use a reinforced concrete wall and utilize it as the spanning element between the two columns. This would be a relatively easy task for the wall, functioning as a deep beam, and would reduce the function of the footing to that of a simple wall footing for the combined wall and column loads.

The following example is a redesign of the foundation in Example 8. It assumes that a full-height wall is not desirable between the two columns, but a short wall can be used below the floor level without a major increase in the footing depth. As shown in Figure 3.44, this short wall is combined with a T-shaped footing.

Example 9 T-Shaped Cantilever Footing With Stiffener

Design data and criteria are the same as in Example 8.

We begin the development of the T-shaped plan by placing a square footing under the interior column, using a size approximately the same as would normally be required if the column were free of the need for combining effects. We then add the connecting footing as a rectangular strip, forming the stem of the T-plan, and also add the stem wall between the two columns. This wall is extended beyond the interior column to stiffen the cantilever as well as the span between the columns.

Referring to Table 3.5, we begin with a 10 ft 6 in. square and 24 in. thick footing under the interior column. The slightly reduced thickness reflects some anticipated alterations. To this we must add sufficient

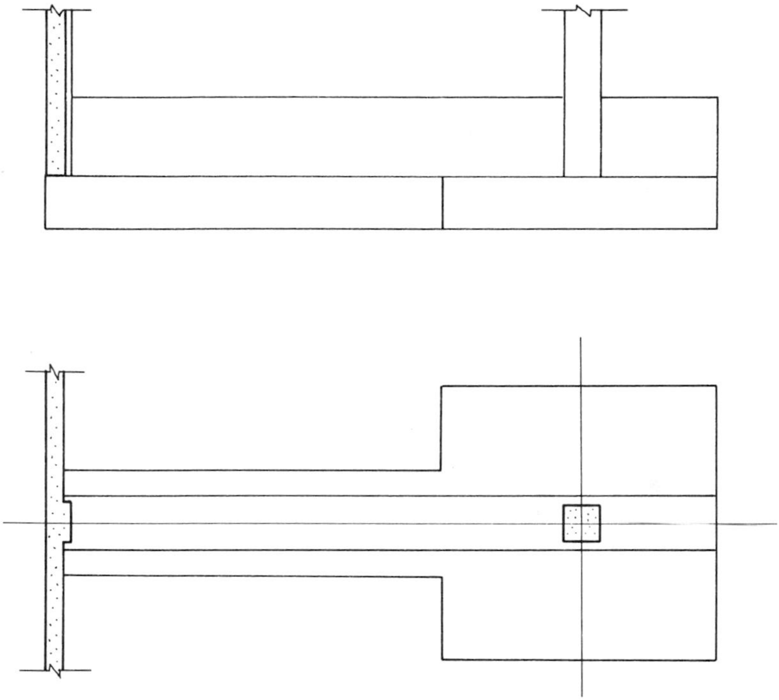

FIGURE 3.44. Use of a T-shaped plan and a stiffening stem wall for the cantilever footing.

additional footing area to keep the total soil pressure under the maximum of 4000 lb/ft^2. If the entire footing is 24 in. thick, the allowable net usable soil pressure for the loads on the footing will be

$$p = 4000 - \tfrac{24}{12}(150) = 4000 - 300 = 3700 \text{ lb/ft}^2$$

Assuming the stem wall to weigh approximately 25 k, the total load on the footing is 625 k, and the required total footing area is thus

$$A = \frac{625{,}000}{3700} = 168.9 \text{ ft}^2$$

Of this required area the 10 ft 6 in. square footing will provide 110.25 ft^2, leaving the stem of the T to provide for the difference of 168.9 - 110.25 = 58.65 ft^2. The stem length, as shown in Figure 3.45,

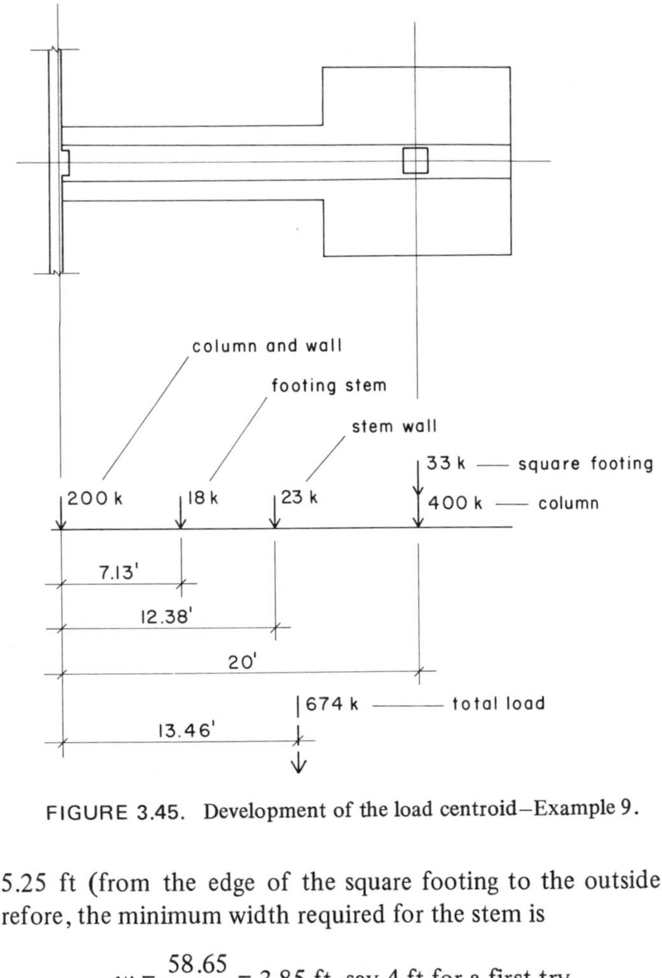

FIGURE 3.45. Development of the load centroid–Example 9.

is 15.25 ft (from the edge of the square footing to the outside end). Therefore, the minimum width required for the stem is

$$w = \frac{58.65}{15.25} = 3.85 \text{ ft, say 4 ft for a first try}$$

Our first guess for the stem wall will be a width of 24 in. and a height above the footing of 36 in. With the sizes of all of the elements determined, we now proceed to find the centroid of the loads. In this case we include the footing weights, since the footing is not uniform in width, as it was in the preceding example. The loads and their locations are shown in Figure 3.45. The calculation for the location of the load centroid is shown in Table 3.11. From this work we find that

TABLE 3.11. Determination of the Location of the Resultant Load (Example 9)

Load (k)	Distance From the Exterior Column (ft)	Moment (k-ft)
200	0	0
18	7.13	128.3
23	12.38	284.7
433	20.00	8660.0
674 ←	Totals	→ 9073.0

$$\bar{y} = \frac{9073.0}{674} = 13.46 \text{ ft}$$

the load centroid is 13.46 ft from the exterior column. This is actually very close to the centroid location in the preceding example, as shown in Figure 3.41.

If we wish to have a uniformly distributed soil pressure, the centroid of the T-shaped footing plan area must be at the same location as the centroid of the loads. The footing areas, considered as analogous loads, are shown in Figure 3.46, and the calculations for determination of the footing centroid are given in Table 3.12. From this work we find that footing centroid is almost 2 ft from the load centroid. Although we intend to modify the footing, it may be useful to calculate the actual soil pressure distribution with the current dimensions in order to demonstrate the effect of the eccentricity. Referring to the general discussion of moment-resistive footings in Section 3.11, we see that as long as there is no potential tension stress on the section, the stress distribution will be determined by the combined stress formula as follows:

$$p = \frac{P}{A} \pm \frac{M}{S}$$

in which P is the total load of 674 k

A is the total footing area of 171.25 ft^2

M is due to the eccentricity of the load, such that

$$M = P \times e = (674)(1.96) = 1321 \text{ k-ft}$$

S is the section modulus of the T about its centroidal axis

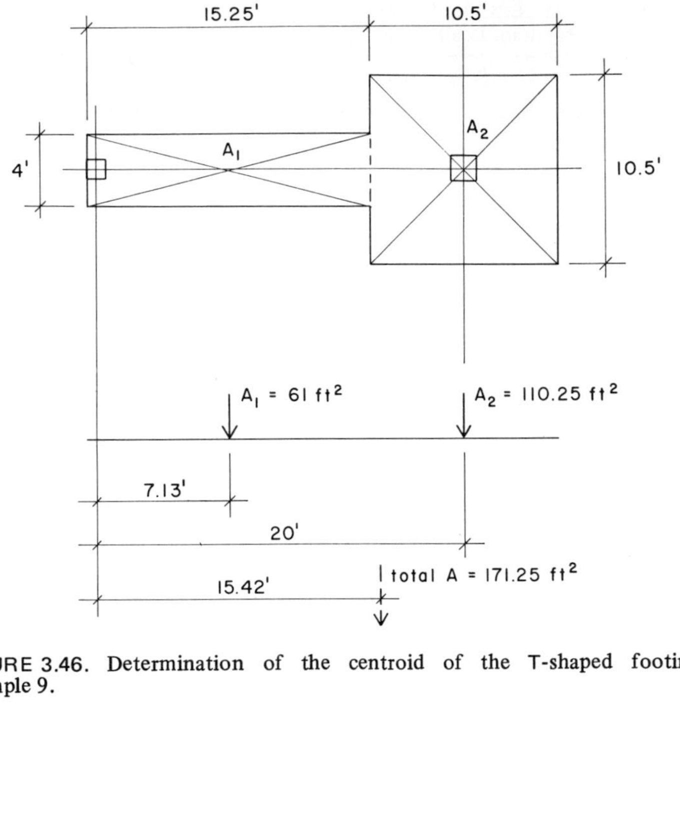

FIGURE 3.46. Determination of the centroid of the T-shaped footing—Example 9.

TABLE 3.12. Determination of the Centroid for the T-Shaped Footing

Area (ft^2)	Distance From the Exterior Column (ft)	Moment (ft^3)
$A_1 =$ 61.00	7.13	435
$A_2 =$ 110.25	20.00	2205
Total: 171.25		2640

$$\bar{y} = \frac{2640}{171.25} = 15.42 \text{ ft}$$

There are actually two S values for the section, since S is equal to I/c and there are two c values for the two different edge distances from the centroid. Figure 3.47 illustrates the determination of the moment of inertia of the T, using the parallel axis relationship for the multiunit section. The calculations for this are given in Table 3.13. As illustrated in Figure 3.48, the stresses thus are

$$p_1 = \frac{P}{A} + \frac{M}{S_1} = \frac{674}{171.25} + \frac{1321}{546}$$

$$= 3.936 + 2.419 = 6.355 \text{ k/ft}^2$$

$$p_2 = \frac{P}{A} - \frac{M}{S_2} = 3.936 - \frac{1321}{886}$$

$$= 3.936 - 1.491 = 2.445 \text{ k/ft}^2$$

Although the maximum stress is clearly in excess of our allowable limit of 4000 lb/ft^2 and is thus not permissible for our example, whether this degree of nonuniformity in the pressure is truly bad is a somewhat more complex judgment. This matter is discussed in the section on moment-resistive footings previously cited.

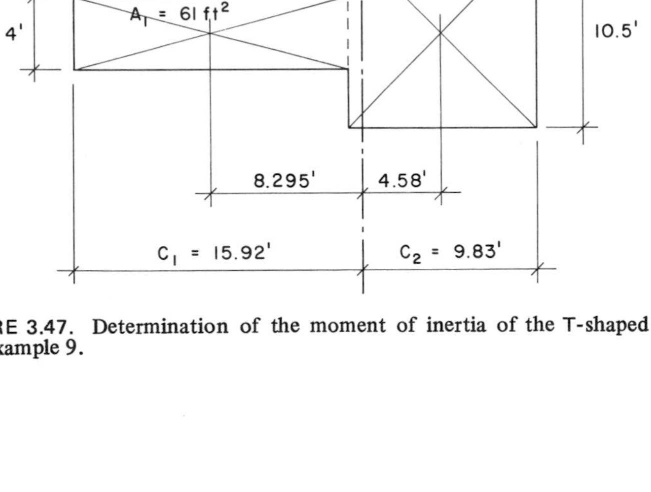

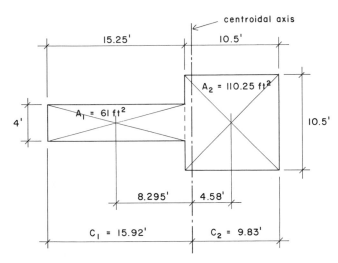

FIGURE 3.47. Determination of the moment of inertia of the T-shaped footing–Example 9.

TABLE 3.13. Determination of the Moment of Inertia of the T-Shaped Footing

Area (ft^2)	$I_0 = \frac{bd^3}{12}$ (ft^4)	d (ft)	Ad^2 (ft^4)	$I_0 + Ad^2$ (ft^4)
A_1 = 61.00	1182	8.295	4197	5379
A_2 = 110.25	1013	4.58	2313	3326
Total moment of inertia for the footing:				8705

$$S_1 = \frac{I}{c_1} = \frac{8705}{15.92} = 546 \text{ ft}^3$$

$$S_2 = \frac{I}{c_2} = \frac{8705}{9.83} = 886 \text{ ft}^3$$

In order to eliminate, or at least to substantially reduce, the eccentricity, we must manipulate the dimensions of the T-shaped footing. To shift the footing centroid in the necessary direction we must shift some of the area from the square portion to the stem portion. Figure 3.49 shows a second version of the footing plan in which we have shifted the areas by trimming one dimension of the square portion to 8 ft and

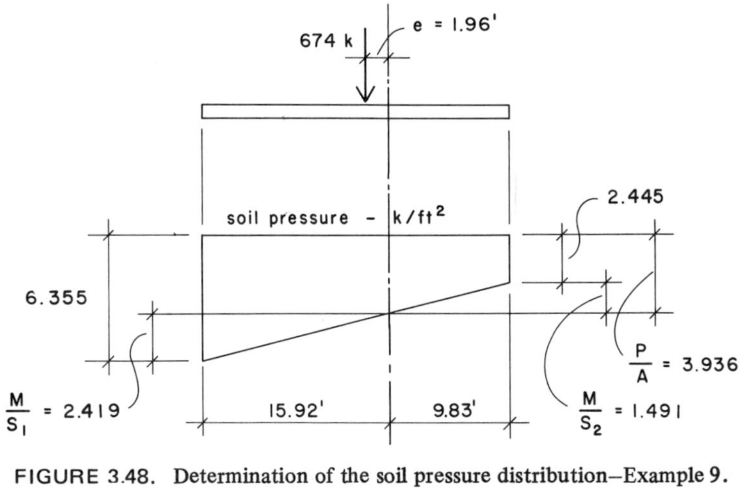

FIGURE 3.48. Determination of the soil pressure distribution–Example 9.

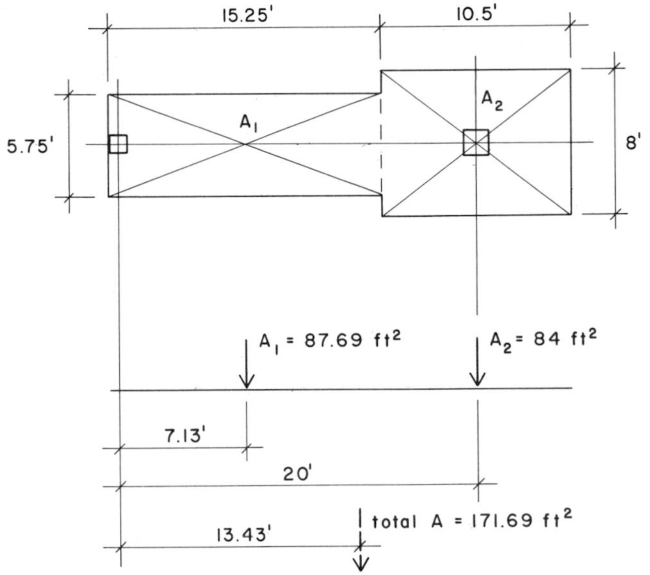

FIGURE 3.49. Determination of the plan centroid for the revised footing–Example 9.

widening the stem to 5 ft 9 in. This requires a recalculation of both the load and footing area centroids. The procedures are the same as those we have illustrated, so we will not show the calculations. The result is to bring the two centroids to within a few inches of each other. With all of the approximations and rough data and dimensioning in this work, it is silly to split hairs any further.

With the very stiff stem wall the two footing areas are reduced to functioning in one-way bending, as for a typical wall footing. They would be essentially designed as such, with major reinforcing provided only in the direction perpendicular to the stem wall. It is also quite possible that the footing thickness could be reduced, since the original estimation was based on a comparison with the square footing required for the interior column. We assume this to be possible and reduce the thickness to 20 in. for the second try, adding the dimension to the stem wall height.

We do not intend to complete the design of this foundation in all respects, but will proceed to investigate the major stresses in the stem wall to determine its general feasibility. The soil pressure for stress investigation of the stem wall will be that due to the column loads only, since the weight of the wall and footing will not cause stress in the wall. Our new plan area has a total of 171.7 ft^2 and the design soil pressure is thus

$$p = \frac{600{,}000}{171.7} = 3494 \text{ lb/ft}^2$$

For the two footing widths this results in linear loading of the stem wall as follows:

$$w_1 = (5.75)(3.494) = 20.1 \text{ k/ft}$$

$$w_2 = (8)(3.494) = 27.95 \text{ k/ft}$$

The loading conditions and the resulting shear and moment diagrams for the stem wall are shown in Figure 3.50. Also shown is the full T-shaped section of the combined wall and footing. For the largest moment, in the span between the columns, this section operates as a T-section with compression in the T-flange (the footing) and tension reinforcing in the top of the wall. For the moment at the interior column the stresses reverse, with compression in the top of the wall. Using an approximate effective d of 56 in., the balanced stress moment capacity of the section ignoring the T action is

$$M_R = Rbd^2 = \frac{(0.226)(24)(56)^2}{12} = 1417 \text{ k-ft}$$

which clearly indicates that concrete flexural stress is not critical for either moment condition.

Because of its considerable depth, the critical section for shear for this beam is quite a distance from the support. The required design shear force is thus less than half of the maximum shear value shown on the diagram in Figure 3.50. With this design shear force the stress in the beam is

$$v = \frac{V}{bd} = \frac{109{,}000}{(24)(56)} = 81.1 \text{ psi}$$

This is somewhat in excess of the limit of 60 psi for the concrete, but could relatively easily be provided for with some vertical reinforcing as shown in Figure 3.50. This reinforcing would likely be provided anyway, so does not constitute a major additional expense.

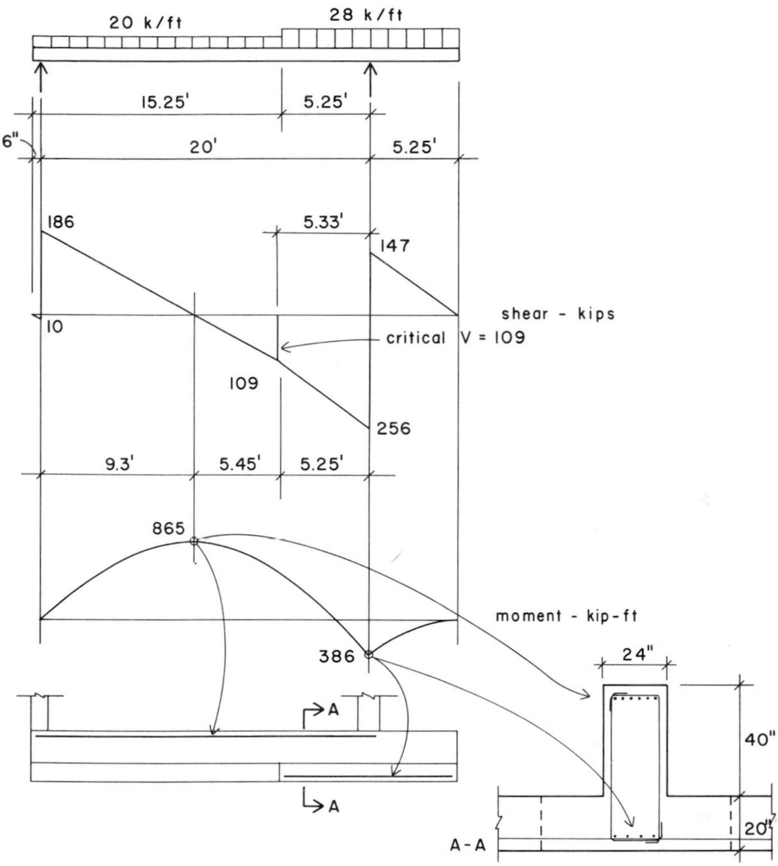

FIGURE 3.50. Beam action of the stiffened cantilever footing–Example 9.

Although the T-section is somewhat more efficient than the simple rectangular section, we will determine approximate values for the tension reinforcing by assuming a *jd* of 0.9 times the effective depth. The area requirements thus are as follows:

At the span between the columns

$$A_s = \frac{865 \times 12}{(20)(0.9)(56)} = 10.30 \text{ in.}^2$$

At the interior column

$$A_s = \frac{386 \times 12}{(20)(0.9)(56)} = 4.60 \text{ in.}^2$$

This is not an unreasonable amount of reinforcing to accommodate in the 24 in. wide beam. Comparison with Example 8 will show that the major reinforcing is considerably less for this foundation. However, the overall advantage of this construction in place of the simple rectangular footing is questionable. The total volume of concrete is only slightly less, and the added costs of forming, the additional required excavation, and the added time for the construction work would likely nullify any savings from use of less concrete and steel. This should not be construed as a general evaluation of the feasibility of such a foundation. In other circumstances, with different column loads, column spacing, and allowable soil pressure, the comparisons may be quite different.

3.9 Special Footings

In many buildings it is necessary to provide special footings for individual elements of the building. Some of the ordinary elements of this type are:

- Stair towers.
- Elevator pits and shaftways.
- Masonry fireplaces and chimneys.
- Large pieces of equipment.
- Large signs, light towers, water towers, and so forth.

There are many different situations in terms of the loads, the materials and details of the construction, and the relation of the elements to the general structural system for the building. In some cases it is best to provide foundation support that is integrated with the general foundation system for the building. In other situations it may be advisable, or required by code, to provide a separate, structurally independent foundation. The following is a discussion of some of the problems that may occur with these foundations.

Stair Towers. In multistory buildings stair walls usually represent part of the permanent unpierced wall system that is vertically continuous through the full building height. They are thus often utilized for bearing walls and for shear walls, in which cases a considerable load may be transferred to their foundations. The stair walls may be provided with individual wall footings when the loads are modest, but a common solution is to provide a single large footing for the entire stair tower, as shown in Figure 3.51. In this case the footing functions as a two-way spanning concrete slab with edge supports. However, since the stairway plan is usually considerably longer in one direction than in the other, it is practical to design the footing as a one-way spanning slab in the short span direction and to rely on the minimum temperature reinforcing in the other direction for the lower stresses.

If the stair walls are of reinforced concrete or masonry and are utilized as both bearing and shear walls, and the tower is several stories high, the load on the foundation may be very significant. If the allowable soil pressure is low, it may be necessary to spread the footing out considerably beyond the periphery of the stair walls, especially if resistance to overturning moments created by lateral loads must be developed. The special problems of shear wall foundations are discussed in Section 3.13.

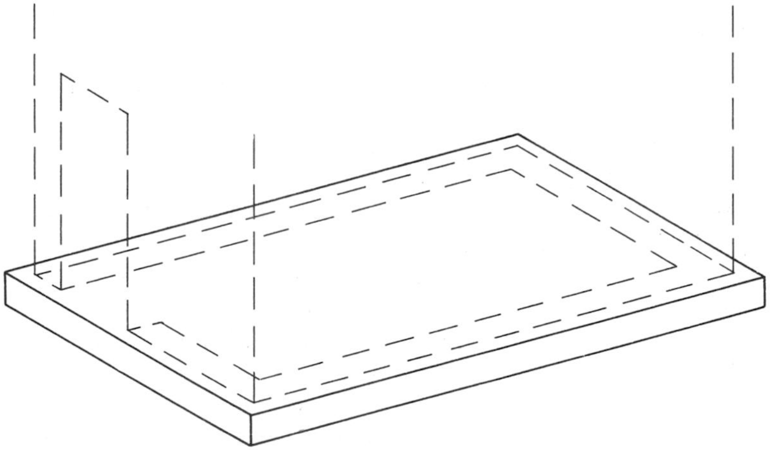

FIGURE 3.51. Typical combined bearing footing for a stair tower wall.

When the stair and stair wall construction is relatively light, and is supported at the separate levels by the building structure, the load on the foundation may consist of nothing more than the weight of the lower stair flight and walls. In this case the foundation may need to be nothing more than a very minimal wall footing.

Elevators. For high-speed elevators with the lift mechanism in the top of the shaft, the foundation must accommodate a pit whose bottom is some distance below the lowest level served. When the elevator serves the basement, this may drop the bottom of the pit somewhat lower than the necessary level for the bottom of footings, which may present a problem if other footings are close to the elevator. The general problem of adjacent footings at different elevations is discussed in Section 3.15.

Hydraulic elevators require the installation of a cylinder that projects a distance into the ground approximately equal to the height of the run of the elevator. The problems of installing the cylinder and the effects it may have on adjacent footings or soil conditions must be carefully studied as part of the design of the elevator system.

The exact requirements of details and dimensions for the base of an elevator system can be established only after the particular manufacturer and particular model of elevator are known. General features and approximate dimensions may be used for preliminary design when the system is an ordinary one, but the final details must relate to a specific elevator.

As with stair towers, the elevator foundation must also support the walls of the shaftway. If these are used as bearing walls or shear walls, the same issues that were discussed for stairs apply.

Fireplaces and Chimneys. Masonry fireplace and chimney construction usually represents a considerable dead load accumulation, although if the plan size is large, the necessary base for the construction of the fireplace or chimney may define a considerable footing area. Thus the actual vertical soil pressure may not be that high when compared with other building loads on smaller footings.

Because of fire separation requirements, this construction may be essentially independent from the rest of the building in a structural sense. Thus the footing provided would be of the nature of one for a free-standing tower. For critical earthquake forces it may be necessary to design the structure for independent resistance to sliding and overturn effects.

Because the load on the soil in this situation is almost all dead load, the potential for settlement must be carefully studied. Some of the possible problems are the following:

Time-dependent settlement. If the supporting soil is a relatively soft clay or other type of soil subject to time-related strain change, the high constant dead-load pressure may cause significant progressive settlement.

Influence on lower soil strata. In a small building the fireplace footing may be considerably larger than the other building footings. Even though design pressures may be low and the dead load pressures equalized for all footings, the larger footings may settle more, as discussed in Section 2.3.

Differential settlement. Although the actual total amount of settlement may be a problem if it is excessive, the more common problem is that of different amounts of settlement by adjacent parts of the building. This can go either way, with the fireplace settling more for reasons mentioned previously, or the adjacent walls or columns settling more because of higher unit soil pressures under their footings. If differential settlement is excessive, it may cause the fracture of roof flashing or floor closure joints or other details of the construction.

Building codes often have a number of detailed requirements for fireplaces and chimneys dealing with both the fire separation and structural problems.

Equipment Bases. Various items of building equipment require a base that must provide both vertical support and anchoring for the equipment. When the equipment is relatively light and sits on a floor slab on fill, it may be adequately supported by the floor slab without special provisions. When the equipment is heavy, however, it may be necessary to provide a stronger support. The two principal means, as shown in Figure 3.52, are to build a strengthened system into the floor construction or to build a separate, isolated foundation for the equipment. The

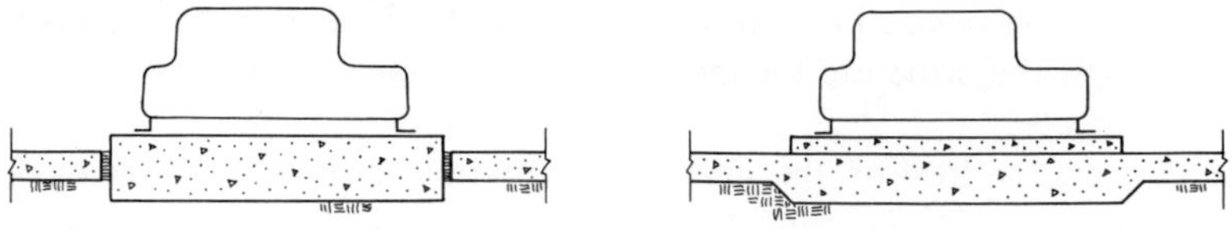

FIGURE 3.52. Options for equipment foundations.

latter solution may be additionally desirable if vibration of the equipment must be isolated from the building structure.

Signs, Towers, and the Like. These items must usually be provided with their own independent foundations, which generally must resist lateral as well as vertical loads. The general problems of moment-resistive footings are discussed in Section 3.11. The various problems of tower foundations are discussed in Section 3.14.

3.10 Columns in Walls

A common problem in the design of building foundations is a foundation wall that must share its location with a row of columns. This happens quite frequently along the exterior walls of buildings with frame structures. In the process of transferring loads to the ground there are various possibilities for the relationships between the columns, the wall, and the foundation elements. Some of the variables to consider in this situation are the following:

Magnitude of the column loads. It is one thing if the column loads are light, as in the case of a one-story building with light construction and short spans. It is quite another thing if the column loads are large, as in the case of a multistory building.

Spacing of the columns. If columns are closely spaced, the idea of using the wall as a spanning or distributing element will have more merit. If columns are quite far apart, it becomes less reasonable to expect the wall to achieve these functions. There is no cut-off dimension for this effect by itself; it must be considered along with the column load, wall height, type of foundation, and so forth.

Height of the wall. For spanning or distributing functions the most critical dimension for the wall is its height. This relates to the relative efficiency of the wall as a beam and the classification of the wall as a spanning member. For the latter relationship, the wall may fall into three categories of behavior as a function of its span-to-depth ratio. As shown in Figure 3.53, these are:

1. An ordinary flexural member (beam) with shear, moment, and deflection as we ordinarily consider it to develop for a beam.
2. A deep beam with virtually no deflection, and with stiffness in proportion to its span that significantly affects the nature of stress and strain on a vertical section in the member.
3. A member so stiff with respect to its span that there is essentially no flexure involved in its action. Instead, it functions in arching, or corbeling, action to bridge the space between supports.

The numerical values shown in Figure 3.53 are approximate limits for identifying these behavioral differences. As the wall changes in depth-to-span ratio there is no sudden switch from one form of action to another, but rather a gradual shift.

Type of foundation. If the foundation consists of deep elements, either piers or groups of piles, the wall is most likely to be designed as a spanning element of one type or another. When the foundation is of the shallow bearing type, there may be several options for the column/wall/footing relationships, as illustrated in the following example.

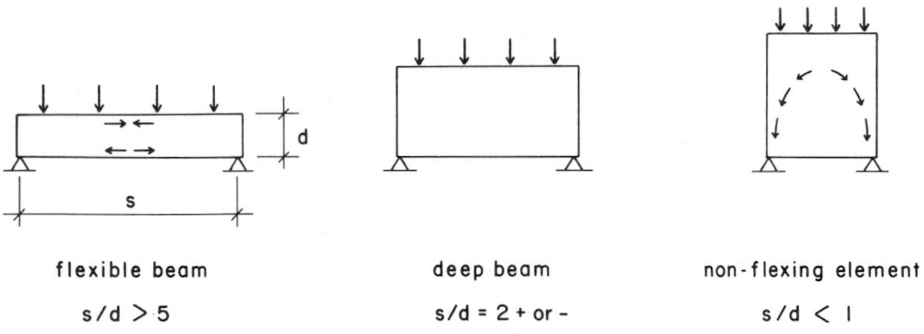

FIGURE 3.53. Effects of the span-to-depth ratio on the action of spanning elements.

Example 10 Foundation Wall With Columns

The situation for this example is shown in Figure 3.54. In addition to the data given in the illustration, we assume the following design criteria:

Allowable soil pressure: 3000 lb/ft^2.
Concrete design strength: $f_c' = 3000$ psi.
Allowable tension on reinforcing: 20,000 psi.

Considering the wall by itself and ignoring the existence of the columns, the minimum wall footing that we obtain from Table 3.2 is 11 in. thick, 42 in. wide. Thus, if we wish to ignore the spanning potential of the wall, this would be the required footing for the wall between the column footings.

For the column, if it was freestanding and not in the line of the wall, the required square footing, from Table 3.5, would be 6 ft square, 16 in. thick. However, in this situation even if it is decided to use a square footing, it would need to be larger, since it also supports part of the wall. And the larger it is, the more of the wall it supports. If we increase its size to 8 ft square, the total load on the footing will be the column load of 100 k plus 8 ft of the wall load at 10 k/ft, or a total of 180 k. This is quite close to the load capacity of the 8 ft square 19 in. thick footing listed in Table 3.5.

This procedure is actually quite commonly used by designers. The required wall footing is used under the wall and a two-way reinforced

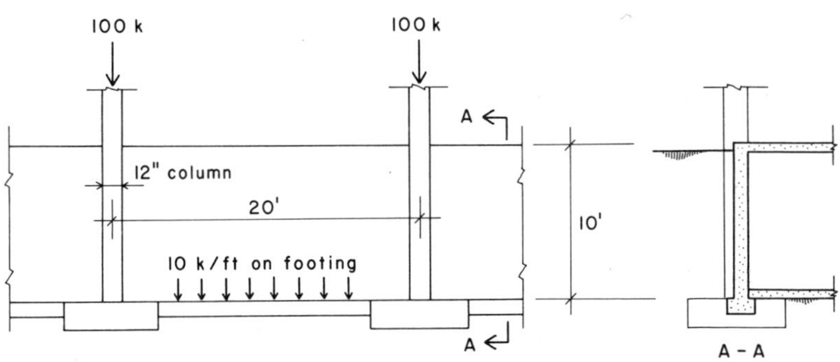

FIGURE 3.54. Foundation wall with columns–Example 10.

square footing is used under the column, ignoring any effects of the wall on the column footing, other than its gravity load. There are, however, several approaches to this problem, as shown in Figure 3.55 and discussed in the following:

1. The approach just described; ignore any structural effects of the wall on the column footing; provide an ordinary wall footing and a square two-way reinforced column footing.
2. The same as Option 1, but consider the wall capable of preventing bending in the column footing in the direction parallel to the wall; design both the wall and column footings for one-way bending.

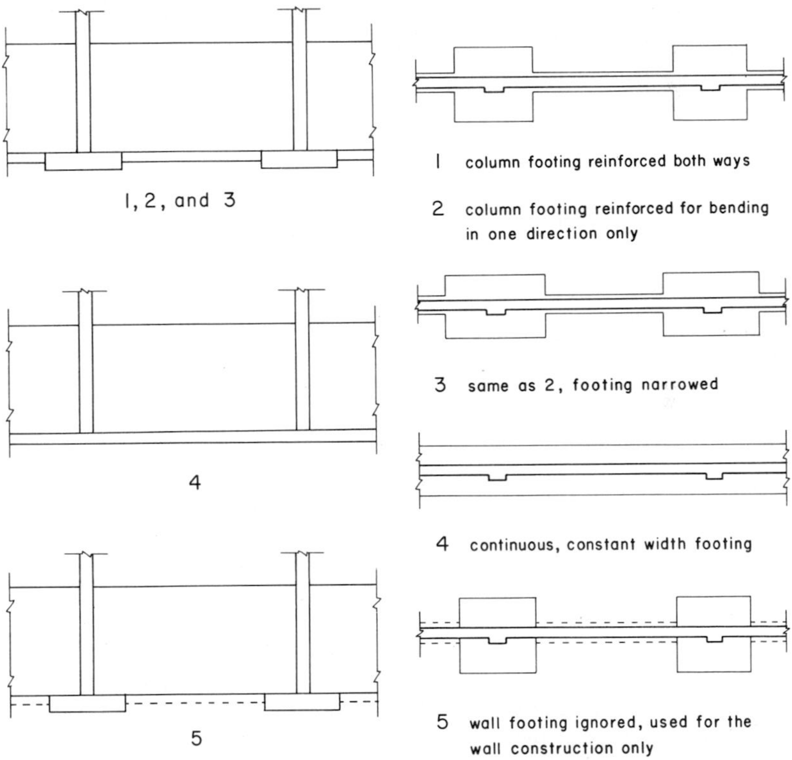

FIGURE 3.55. Options for the development of foundations for walls with columns.

3. A better version of Option 2; make the column footing slightly rectangular, which shortens the cantilever for the one-way bending.
4. The ultimate extension of Option 3; consider the wall to function as a distributing grade beam and blend the wall and column footings into a continuous constant width footing. Design the wall to span from center-to-center of the columns for the uniform upward load consisting of the net difference between the footing pressure and the load on top of the footing (5 k/ft in the example).
5. The opposite of (4); provide a column footing capable of supporting the entire combined loads on the column and wall. Provide a minimal platform footing under the wall, but ignore its bearing capacity. Design the wall as a continuous beam spanning from column-to-column and carrying the wall loading of 10 k/ft.

A real comparison of the relative merits of these options can be made only by designing all of them. This may involve more than five designs, since there are several possible variations for Options 3 and 5. It is doubtful that any designer would do that much work to optimize the design, although with some experience fast approximate designs would not be that difficult to perform. However, experience would also permit some judgment about which the really feasible options are. In addition, there may quite likely be some additional considerations regarding problems of the excavation, the general building construction, and other structural functions of the wall.

With facts given in this example, we would judge that Option 3 or Option 4 would provide the best design. If Option 4 is used, the total load on the continuous width wall footing becomes

$$w = 10 + \frac{100}{20} = 10 + 5 = 15 \text{ k/ft}$$

From Table 3.2 this would require a footing approximately 5 ft 6 in. wide and 16 in. thick. A comparison of this footing with the first design that totally ignores the spanning capability of the wall shows that the concrete volume is almost the same in the two designs, but the weight of reinforcing bars in the continuous width footing is almost one third of the total in the wall and column footings of the previous design. For a fair comparison, however, we must include the extra reinforcing required in the wall for the spanning function.

With a span-to-depth ratio of two, the wall qualifies as a deep beam, and should be designed as such. We do not recommend the use of working stress design methods for such a situation, but suggest rather that deep beams be designed by strength design methods using the requirements in the latest edition of the ACI code (Ref. 10) for design criteria. We may, however, use some approximate analyses to estimate the requirements for the wall.

As shown in Figure 3.56, the wall functions as a continuous beam spanning from column to column. As discussed previously, the load on this beam is the net difference between the total soil pressure on the bottom of the footing (acting upwards) and the load on the wall plus the wall weight (acting down), or

$$w = 15 - 10 = 5 \text{ k/ft}$$

The two primary concerns for the wall are the shear stress and the requirements for tension reinforcing in the top and bottom of the wall. For a really conservative analysis we may take the shear force at the end of the span and use it to find a shear stress in the wall. This force, as shown in Figure 3.56, is

$$V = \frac{wL}{2} = \frac{(5)(20)}{2} = 50 \text{ k}$$

and, assuming the 10 ft high wall to have an approximate d of 115 in. and a minimum thickness of 10 in.

$$v = \frac{V}{bd} = \frac{50{,}000}{(10)(115)} = 43.5 \text{ psi}$$

This is a shear stress below any limits mentioned for the concrete with f_c' of 3000 psi. Furthermore, the wall will have some shear reinforcing in the form of the usual vertical and horizontal bars provided. Furthermore, not even in deep beam analysis is it required to use the full end span shear force, so the wall would seem not to be critical for shear stress.

For the continuous beam flexure analysis we may also reasonably conservatively use the maximum moment at the supports as $0.10\ wL^2$, and at midspan as $0.08\ wL^2$. The moment at the support will be used to find the reinforcing required in the bottom of the wall and the moment at midspan will be used to find the reinforcing in the top of

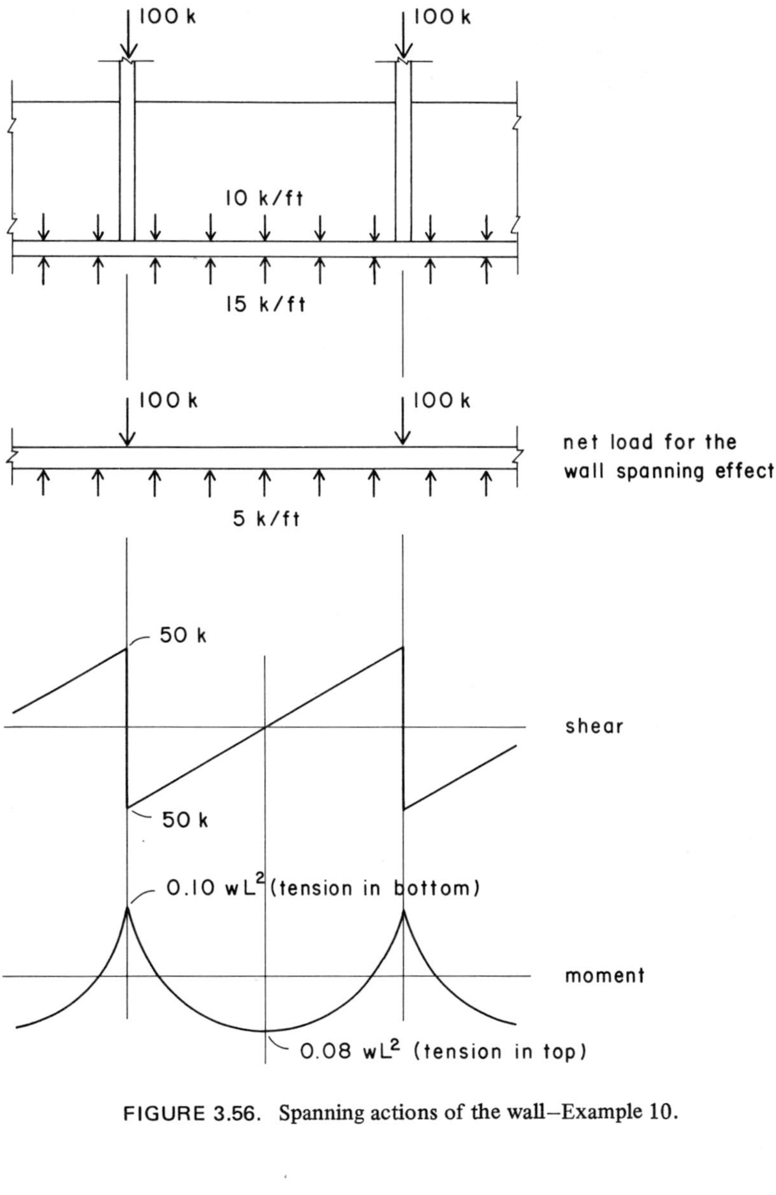

FIGURE 3.56. Spanning actions of the wall–Example 10.

the wall. This is the opposite of our usual situation for continuous beams, but it must be remembered that the net load on the beam is upward. These moments and corresponding area requirements are as follows:

At the support:

$$M = 0.10\,wL^2 = (0.10)(5)(20)^2 = 200 \text{ k-ft}$$

$$A_s = \frac{M}{f_s jd} = \frac{(200)(12)}{(20)(0.9)(115)} = 1.16 \text{ in.}^2$$

At midspan:

$$M = 0.08\,wL^2 = (0.08)(5)(20)^2 = 160 \text{ k-ft}$$

$$A_s = \tfrac{8}{10}(1.16) = 0.93 \text{ in.}^2$$

These requirements could be satisfied by placing a single large-diameter bar at each location, in this case a No. 10 bar in the bottom of the wall and a No. 9 bar in the top. However, we recommend the use of several small bars at each location, rather than the single large bar. This may be done most simply by using the usual horizontal wall reinforcing bars and merely grouping some of them in the top and bottom of the wall. With a No. 5 bar, for example, this would require a group of three in the top of the wall and four in the bottom. If this is done, we would consider that we have actually added only a total of five bars, since one bar would normally be provided at each location.

With the addition of the wall reinforcing just determined, the total weight of steel for the spanning wall and continuous width footing still comes up to only about half of that required for the design that ignores the wall spanning function.

The reader is cautioned not to draw too many general conclusions from this example. There are several facts that qualify the data in this example. In other situations they may shift the choice of foundation system to one of the other options mentioned. In our example some of the important specific facts are the following:

1. The wall is reasonably deep with respect to the span. This makes it relatively quite efficient for the flexural functions and results in modest area requirements for the top and bottom tension rein-

forcing. It also results in virtually no measurable deflection of the spanning wall/beam, which tends to assure that the soil pressure on the continuous-width footing will be constant as assumed.

2. The load on the wall is relatively high in comparison to the load on the columns. This tends to make the net load for the spanning wall/beam lower, further increasing the feasibility of the design option of the continuous-width footing. If the load on the wall in our example was only 5 k/ft and the load on the column was 200 k, the total load on the footing would be the same, but the net load for the spanning wall/beam would be doubled.
3. With the relatively small column load of 100 k the column would likely be reinforced with small-diameter bars, making it possible to develop the required dowel length in the thinner continuous-width footing. With a larger column load and large diameter bars, the thicker column footing may be desirable to develop the dowels.

When the column load is quite high but the wall is still judged to have some spanning potential, it may be reasonable to consider the option in which the column is provided with an ordinary square footing and the wall is designed to span from footing to footing.

3.11 Moment-Resistive Footings

Bearing footings must occasionally resist moments, in addition to some combination of vertical and horizontal forces. Some of the situations that produce this effect, as shown in Figure 3.57, are the following:

Freestanding walls. When a wall is supported only at its base and must resist horizontal forces on the wall, it requires a moment-resistive foundation. Examples are exterior walls used as fences and interior walls that are not full story in height. The horizontal forces are usually due to wind or seismic effects.

Cantilever retaining walls. Cantilever retaining walls are essentially freestanding walls that must sustain horizontal earth pressures. The various aspects of behavior and the problems of design of such walls are discussed in Section 5.3.

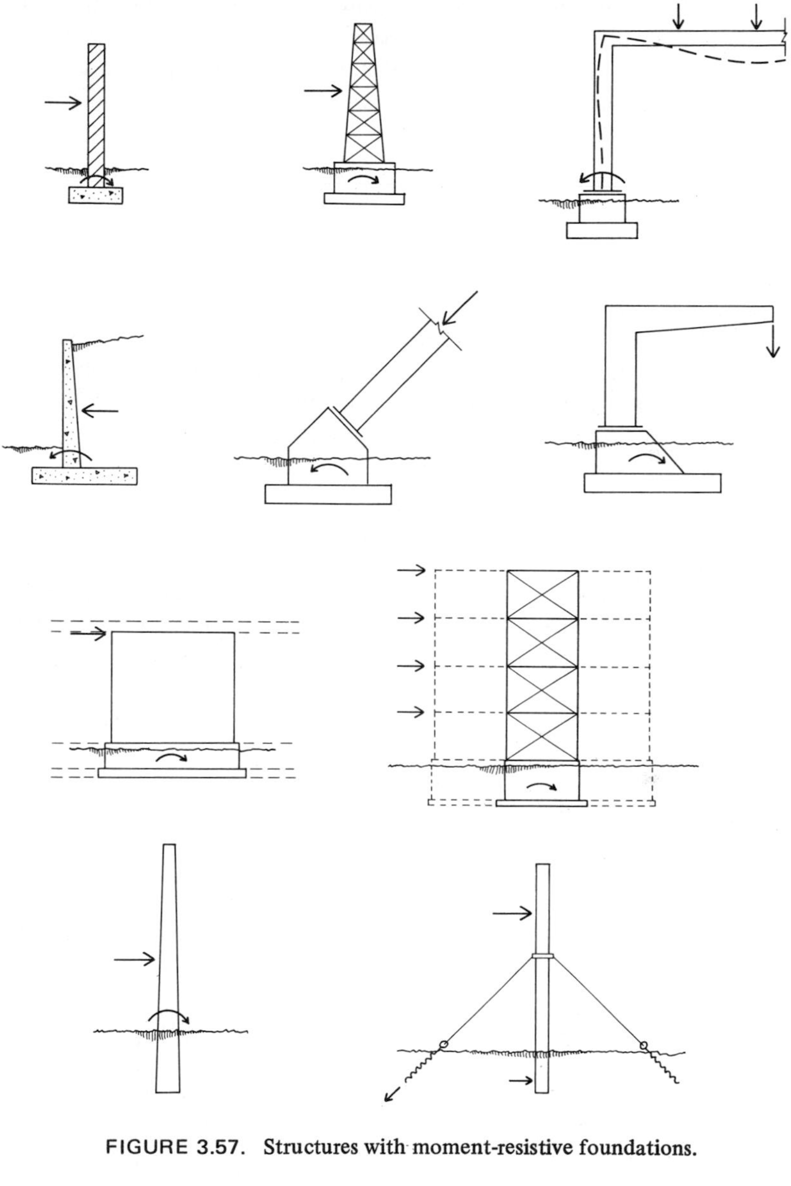

FIGURE 3.57. Structures with moment-resistive foundations.

Bases for shear walls and trussed frames. The overturning effect at the bottom of a shear wall must be resisted by the foundation. When the wall is relatively isolated in plan, as in the case of some interior walls, the foundation for the wall may be developed in a manner similar to that for a freestanding tower.

Supports for rigid frames, arches, cable structures, etc. The foundations for these types of structures must often sustain horizontal forces and moments, even for vertical gravity loading. The special problems of abutments are discussed in Section 5.4.

Bases for chimneys, signs, towers, flagpoles, etc. Any freestanding vertical element supported only at its base must have a moment-resistive foundation. Such a foundation may be quite simple and modest when the element is small, or may be a major engineering undertaking when the element is very tall and the horizontal forces are very high.

Figure 3.58 shows a situation in which a simple rectangular footing is subjected to forces that require the resistance of vertical force, horizontal sliding, and overturning moment. The development of resistance to the horizontal force is discussed in Chapter 5. Our concern here is for the combined effects of the vertical force and overturning moment

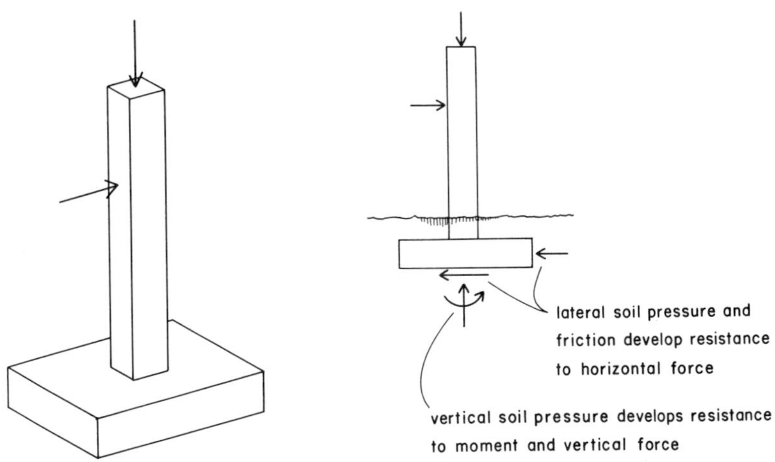
lateral soil pressure and
friction develop resistance
to horizontal force
vertical soil pressure develops resistance
to moment and vertical force

FIGURE 3.58. Typical force development in moment-resistive foundations.

and the resultant combination of vertical soil pressures that they develop.

Figure 3.59 illustrates our usual approach to the combined direct force and moment on a cross section. In this case the "cross section" is the contact face of the footing with the soil. However the combined

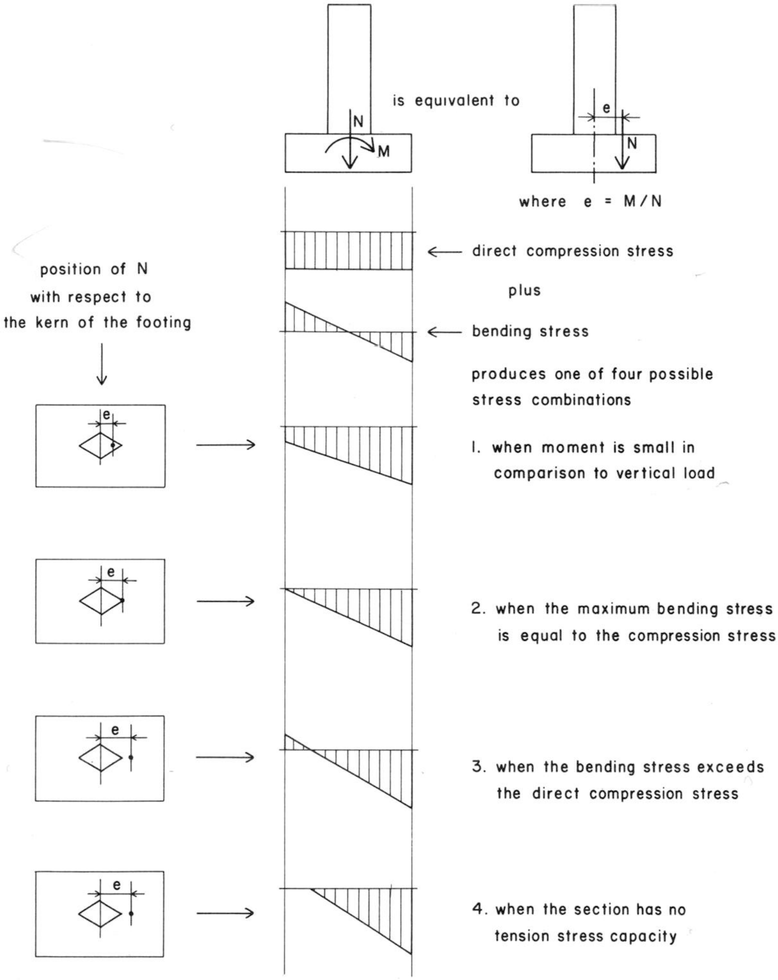

FIGURE 3.59. Analysis for stress due to combined compression and moment.

force and moment may originate, we make a transformation into an equivalent eccentric force that produces the same effects on the cross section. The direction and magnitude of this mythical equivalent *e* are related to properties of the cross section in order to qualify the nature of the stress combination. The value of *e* is established by simply dividing the force normal to the cross section by the moment, as shown in the figure. The net, or combined, stress distribution on the section is visualized as the sum of the separate stresses due to the normal force and the moment. For the stresses on the two extreme edges of the footing the general formula for the combined stress is

$$p = \frac{N}{A} \pm \frac{Nec}{I}$$

We observe three cases for the stress combination obtained from this formula, as shown in the figure. The first case occurs when *e* is small, resulting in very little bending stress. The section is thus subjected to all compressive stress, varying from a maximum value on one edge to a minimum on the opposite edge.

The second case occurs when the two stress components are equal, so that the minimum stress becomes zero. This is the boundary condition between the first and third cases, since any increase in the eccentricity will tend to produce some tension stress on the section. This is a significant limit for the footing since tension stress is not possible for the soil-to footing contact face. Thus Case 3 is possible only in a beam or column where tension stress can be developed. The value of *e* that corresponds to Case 2 can be derived by equating the two components of the stress formula as follows:

$$\frac{N}{A} = \frac{Nec}{I}, \qquad e = \frac{I}{Ac}$$

This value for *e* establishes what is called the kern limit of the section. The kern is a zone around the centroid of the section within which an eccentric force will not cause tension on the section. The form of this zone may be established for any shape of cross section by application of the formula derived for the kern limit. The forms of the kern zones for three common shapes of section are shown in Figure 3.60.

When tension stress is not possible, eccentricities beyond the kern

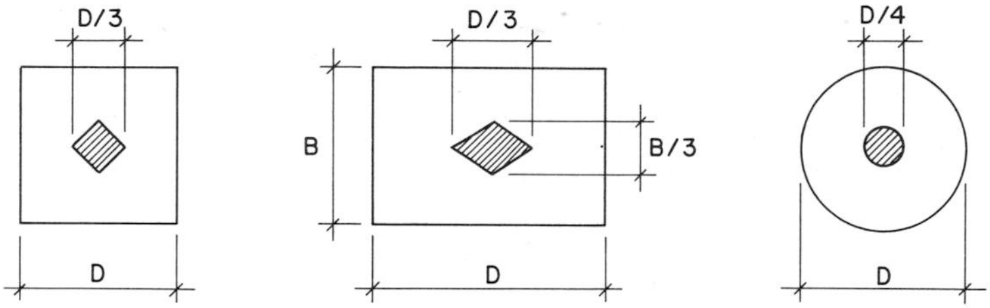

FIGURE 3.60. Kern limits for common shapes.

limit will produce a so-called cracked section, which is shown as Case 4 in Figure 3.59. In this situation some portion of the section becomes unstressed, or cracked, and the compressive stress on the remainder of the section must develop the entire resistance to the force and moment.

Figure 3.61 shows a technique for the analysis of the cracked section, called the pressure wedge method. The pressure wedge represents the total compressive force developed by the soil pressure. Analysis of the static equilibrium of this wedge and the force and moment on the section produces two relationships that may be utilized to establish the dimensions of the stress wedge. These relationships are:

1. The total volume of the wedge is equal to the vertical force on the section. (Sum of the vertical forces equals zero.)
2. The centroid of the wedge is located on a vertical line with the force on the section. (Sum of the moments on the section equals zero.)

Referring to Figure 3.61, the three dimensions of the stress wedge are w, the width of the footing; p, the maximum soil pressure; and x, the limit of the uncracked portion of the section. With w known, the solution of the wedge analysis consists of determining values for p and x. For the rectangular footing, the simple triangular stress wedge will have its centroid at the third point of the triangle. As shown in the figure, this means that x will be three times the dimension a. With the value for e determined, a may be found and the value of x established.

The volume of the stress wedge may be expressed in terms of its three dimensions as follows:

$$V = \tfrac{1}{2} wpx$$

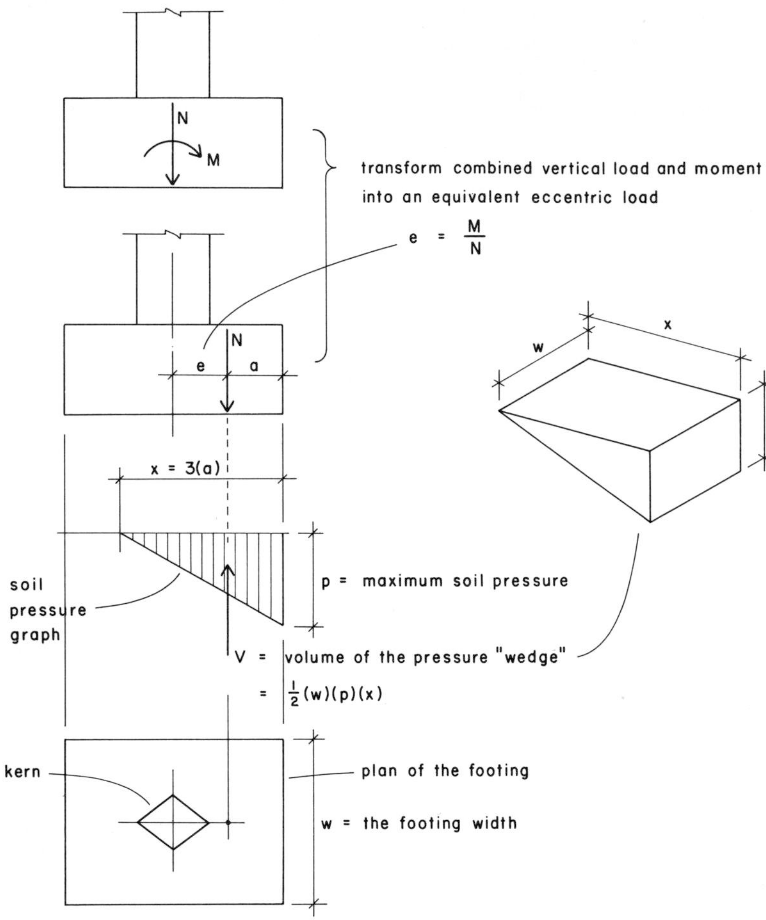

FIGURE 3.61. Analysis of the cracked section by the pressure wedge method.

Using the static equilibrium relationship stated previously, this volume may be equated to the force on the section. Then, with the values of w and x established, the value for p may be found as follows:

$$N = V = \tfrac{1}{2}\, wpx$$

$$p = \frac{2N}{wx}$$

All four cases of combined stress shown in Figure 3.61 will cause rotation of the footing due to deformation of the soil. The extent of this rotation and the concern for its effect on the supported structure must be considered carefully in the design of the footing. It is generally desirable that long-term loads (such as dead loads) not develop uneven stress on the footing. This is especially true when the soil is highly deformable or is subject to long-term continued deformation, as is the case with soft, wet clay. Thus it is preferred that stress conditions as shown for cases 2 or 4 in Figure 3.61 be developed only with short-term live loads.

When foundations have significant depth below the ground surface other forces will develop to resist moment, in addition to the vertical

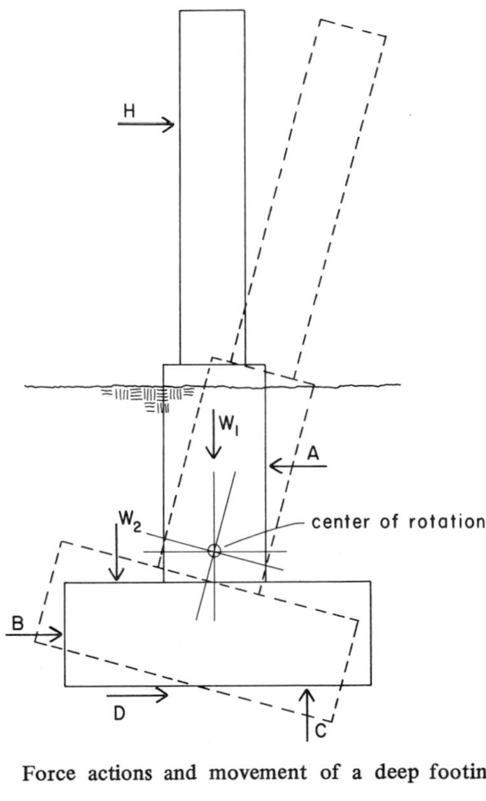

FIGURE 3.62. Force actions and movement of a deep footing subjected to moment.

pressure on the bottom of the footing. Figure 3.62 shows the general case for such a foundation. The moment effect of the horizontal force is assumed to develop a rotation of the foundation at some point between the ground surface and the bottom of the footing. The position of the rotated structure is shown by the dashed outline. Resistance to this movement is visualized in terms of the three major soil pressure effects plus the friction on the bottom of the footing.

When the foundation is quite shallow, as shown in Figure 3.63, the rotation point for the foundation moves down and toward the toe of the footing. It is common in this case to assume the rotation point to be at the toe, and the overturning effect to be resisted only by the weights of the structure, the foundation, and the soil on top of the footing. Resisting Force A in this case is considered to function only in

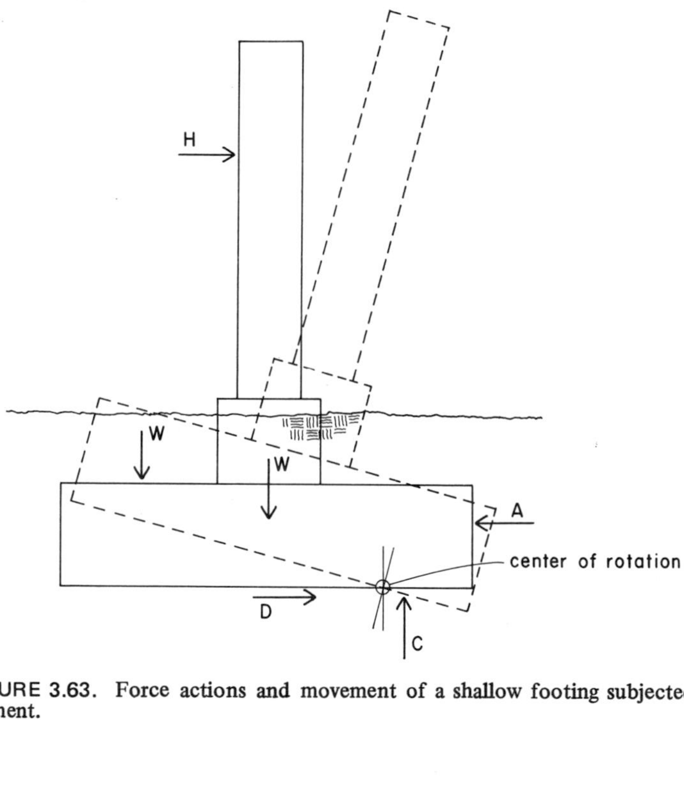

FIGURE 3.63. Force actions and movement of a shallow footing subjected to moment.

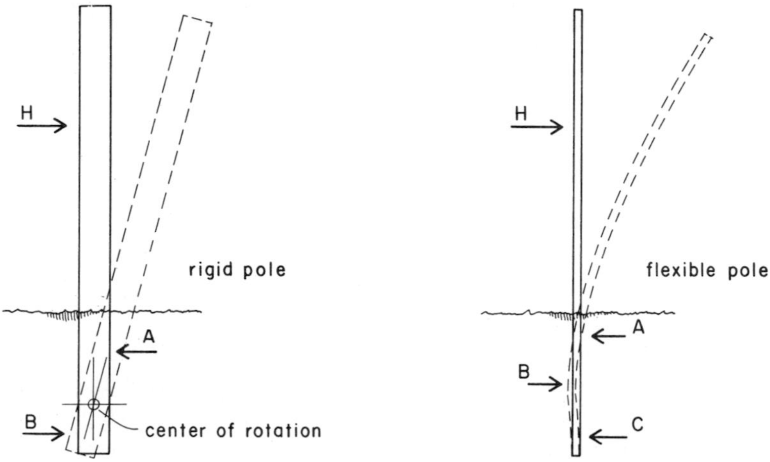

FIGURE 3.64. Force actions and movements of pole-type structures subjected to lateral loads.

assisting the friction to develop resistance to the horizontal force in direct force action.

When a foundation is very deep and is essentially without a footing, as in the case of a pole, resistance to moment must be developed entirely by the forces A and B, as shown in Figure 3.64. If the structure is quite flexible, its bending will cause the two forces to develop quite close to the ground surface, making the extension of the element into the ground beyond this point of little use in developing resistance to moment.

3.12 Footings for Freestanding Walls

Foundations for freestanding walls are usually quite shallow. When the walls are inside the building, frost protection is usually not a problem. When they occur outside and are not connected to the building it is usually not considered necessary to extend them below the frost line. This is a matter of judgment, however, and may in some cases be restricted by local building codes.

The following example illustrates the design for an exterior masonry wall with a shallow footing.

EXAMPLE 11 FOOTING FOR A FREESTANDING WALL

The wall and its footing are as shown in Figure 3.65. Design data and criteria are as follows:

Wall: 8 in. concrete block; weight = 60 lb/ft^2 of wall surface.
Maximum soil pressure: 1000 lb/ft^2.
Wind load: 10 lb/ft^2, horizontal on the wall surface.

Part of the design concerns the stress analysis of the masonry and the adequacy of the anchorage provided by the doweled reinforcing. These must resist the horizontal force and overturn at the top of the footing. We assume these to be adequate in this case and will proceed to investigate the behavior of the foundation.

The three situations to be analyzed are as follows:

1. The maximum soil pressure on the bottom of the footing.
2. The safety factor against sliding, with resistance to sliding devel-

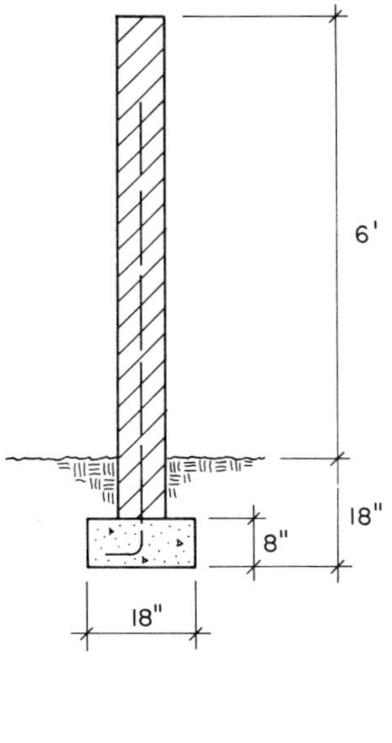

FIGURE 3.65. Footing for a freestanding wall—Example 11.

oped by friction on the bottom of the footing and passive soil pressure against the face of the footing and the wall.

3. The safety factor against overturning.

As in many other situations, the process for designing this footing consists of making some assumptions and guesses to establish enough information to be able to perform the necessary analyses to test the adequacy of the first try. The three facts necessary for the analysis just described are the footing width, the footing thickness, and the depth of the footing below the ground surface. However arrived at, the dimensions given in Figure 3.65 are our first estimates.

Figure 3.66 shows the various forces involved in the three analyses previously described. As labeled in the figure, these are:

H: The total horizontal wind force whose resultant acts at the midheight of the aboveground portion of the wall.

$$H = (10\ \text{lb/ft}^2)(6\ \text{ft}) = 60\ \text{lb}$$

W_1: The weight of the wall.

$$W_1 = (60\ \text{lb/ft}^2)(6.83\ \text{ft}) = 410\ \text{lb}$$

W_2: The weight of the footing.

$$W_2 = (0.67\ \text{ft})(1.5\ \text{ft})(150\ \text{lb/ft}^3) = 150\ \text{lb}$$

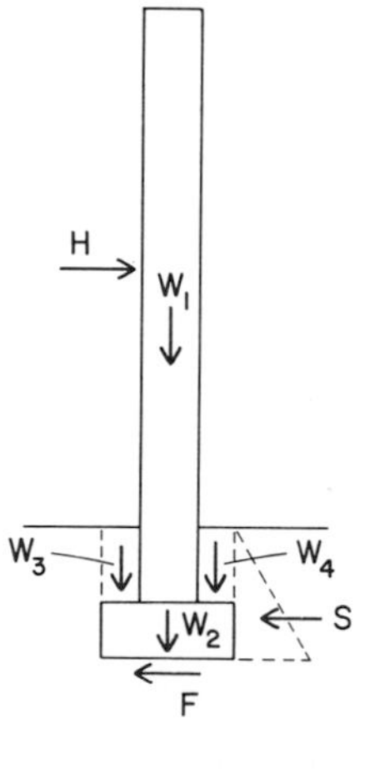

FIGURE 3.66. Active and resistive forces on the freestanding wall—Example 11.

W_3 and W_4: The weight of the soil on top of the footing (assumed to be 80 lb/ft³).

$$W_3 = W_4 = (\tfrac{5}{12}\text{ ft})(0.83\text{ ft})(80\text{ lb/ft}^3) = 28\text{ lb}$$

F: The friction on the bottom of the footing, as developed by the gravity loads.

S: The passive resistance of the soil to horizontal pressure.

For the overturning analysis we consider only the horizontal wind force and the resistance due to the gravity loads. The rotation point for the overturning moment is assumed to be at the toe of the footing. The calculations for the determination of the overturning moment and the restoring moment are given in Table 3.14. Since the safety factor against overturning is found to be in excess of the usual requirement of 1.5, the footing is adequate in this regard.

Friction and passive resistance depend on the type of soil. Criteria vary from reference to reference, so the first source to consider is the applicable building code. We will utilize the data from Table 29-B of the *Uniform Building Code* (see the appendix) in which soil conditions are generalized into five categories. For the relatively low allowable soil pressure in our example we will assume the soil to be in one of the last two groups in the table and will analyze for both to illustrate the process.

TABLE 3.14. Overturn Analysis for the Freestanding Wall (Example 11)

Type of Force	Force (lb)	Moment Arm (in.)	Overturning Moment (lb-in.)	Restoring Moment (lb-in.)
H	60	54	3240	
W_1	410	9		3690
W_2	150	9		1350
W_3	28	15.5		434
W_4	28	2.5		70
Totals	616 lb		3240 lb-in.	5544 lb-in.

$$\text{Safety factor: } \frac{\text{restoring moment}}{\text{overturning moment}} = \frac{5544}{3240} = 1.71$$

For the cohesionless and generally low-plasticity soils in Group 4, the table gives the following design values:

Allowable bearing pressure: 1500 lb/ft^2.
Passive resistance: 150 lb/ft^2 per foot of depth.
Sliding resistance: 0.25 times the dead load.

The total sliding resistance, force F in Figure 3.66, is thus

$$F = (0.25)(616) = 154 \text{ lb}$$

The total passive resistive force, S in Figure 3.66, is determined as the area of the triangular stress graph shown in dotted outline in the figure. Using the value from the table, the maximum stress is determined to be

$$p = (1.5 \text{ ft})(150) = 225 \text{ lb/ft}^2$$

and the total resistance is

$$S = \tfrac{1}{2}(1.5)(225) = 169 \text{ lb}$$

The safety factor in this analysis is already included in the table values, which are given as *allowable* values rather than *ultimate* values. We therefore simply compare the sum of F and S with the horizontal wind force H, and observe that sliding is not a critical concern.

For the high clay content soils in Group 5, the *Uniform Building Code* table gives a passive resistance value of 100 lb/ft^2 per foot of depth. Thus the value of the S force would be two thirds of that calculated previously, or approximately 113 lb. For sliding with this soil group the table gives a value based on the contact area of the footing, rather than a coefficient of friction based on the dead load. Using this we calculate the value for S as

$$S = (1.5 \text{ ft}^2)(130 \text{ lb/ft}^2) = 195 \text{ lb}$$

Note that the footnote to the table limits this force to a maximum of one half the total dead load, which is not a critical consideration for our example.

For this soil group the sum of the resistive forces is also clearly in excess of the wind force. Although it is redundant to consider it for our example, the resistive forces could be increased by one third since the load is due to wind.

For the vertical soil pressure on the bottom of the footing we consider the combined effects of the wind and gravity forces. Since the gravity loads are symmetrical on the footing, they contribute only to the vertical force in this case. The combined effect on the footing is thus as follows:

$$N = 616 \text{ lb (from Table 3.14)}$$

$$M = 3240 \text{ lb-in. (from Table 3.14)}$$

and the equivalent eccentricity, as discussed in Section 3.11, is

$$e = \frac{M}{N} = \frac{3240}{616} = 5.26 \text{ in.}$$

For the rectangular section of the footing contact face the kern limit is 18/6 = 3 in., which indicates that the soil stress condition is Case 4, as shown in Figure 3.59. For the pressure wedge analysis the maximum soil pressure, as derived in Section 3.11, is determined as

$$p = \frac{2N}{wx}$$

in which: N is the total vertical force of 594 lb.
w is the other dimension of the footing "section," or 12 in.
x is three times the distance of the equivalent eccentric force from the edge of the footing, or as shown in Figure 3.67, $3(9 - 5.26) = 3(3.74) = 11.22$ in.

Applying these to the formula, the maximum soil pressure is thus

$$p = \frac{2N}{wx} = \frac{(2)(616)}{(12)(11.22)} = 9.150 \text{ psi}$$

$$= (9.150)(144) = 1318 \text{ lb/ft}^2$$

This soil pressure should be compared to a design value based on a one third increase for wind load. Thus the limit for the soil stress is

$$p = (1.333)(1000) = 1333 \text{ lb/ft}^2$$

This indicates that the soil pressure is approximately at the critical level. However, some designers would prefer that the stress distribution on the footing not be permitted to develop the so-called cracked section, that is, that the eccentricity not exceed the kern limit for the

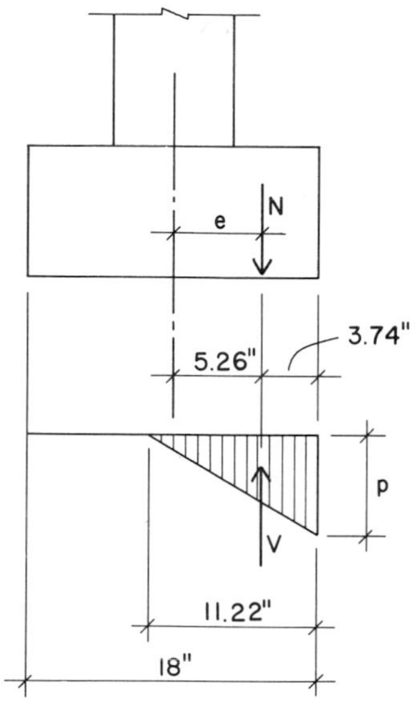

FIGURE 3.67. Analysis of soil pressure for the footing–Example 11.

section. One argument for this is that the high stress concentration on one side of the footing will produce considerable soil deformation at the edges of the footing. With repeated application of horizontal forces, in both directions, weakened resistance to rocking effects will be developed, as shown in Figure 3.68.

If this argument is accepted, it would be necessary to increase the width of the footing in this example by a considerable amount, since the calculated eccentricity of 5.26 in. greatly exceeds the kern limit of 3 in. If the dimension of 5.26 in. is established as the kern limit, the required footing width would be

$$(6)(5.26) = 31.6 \text{ in.}$$

Since the added footing size will increase the dead load, the revised footing width could be slightly less than this. If we change the footing width to 27 in., the new dead load will be 741 lb and the new eccentricity will be

$$e = \frac{3240}{741} = 4.37 \text{ in.}$$

which is within the kern limit of 4.5 in. for the 27 in. width.

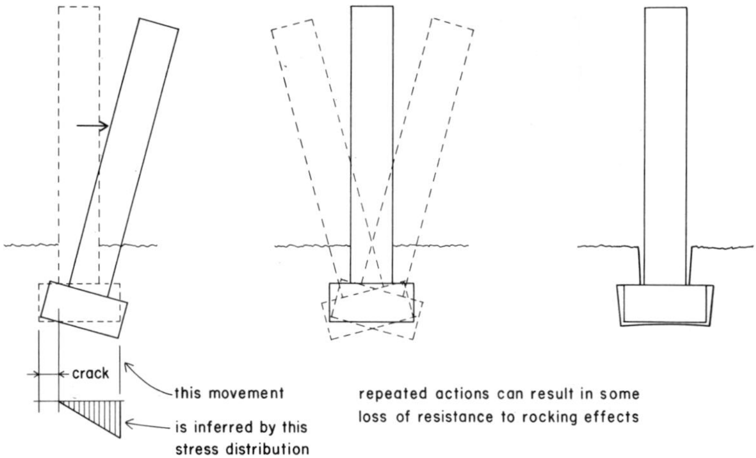

FIGURE 3.68. Effects of repeated lateral loading on freestanding walls.

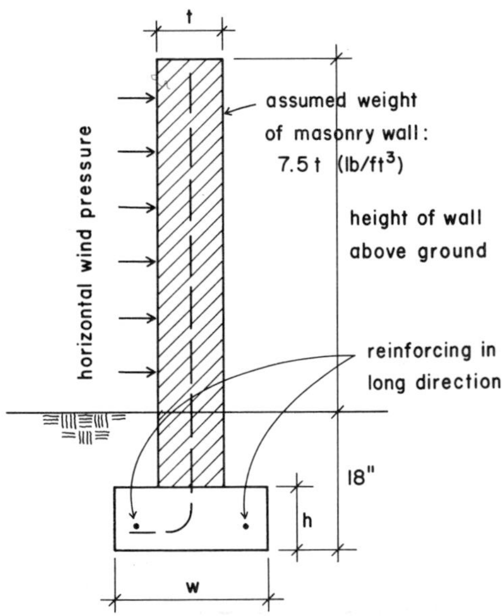

Reference Figure for Table 3.15.

TABLE 3.15. Footings for Freestanding Walls

Wall Height Above Ground (ft)	Horizontal Wind Pressure on Exposed Wall Surface											
	$10\ lb/ft^2$				$20\ lb/ft^2$				$30\ lb/ft^2$			
	t (in.)	h (in.)	w (in.)	Reinf.	t (in.)	h (in.)	w (in.)	Reinf.	t (in.)	h (in.)	w (in.)	Reinf.
4	6	6	15	2 No. 3	6	8	26	2 No. 4	6	8	36	2 No. 5
5	6	6	18	2 No. 3	8	8	31	2 No. 4	8	10	41	3 No. 4
6	8	8	22	2 No. 4	8	10	36	3 No. 4	10	12	45	3 No. 5
7	8	8	25	2 No. 4	10	10	40	3 No. 4	10	12	52	3 No. 5
8	10	10	26	2 No. 4	10	12	44	3 No. 5	12	12	56	4 No. 5
9	10	10	29	2 No. 5	12	12	46	3 No. 5	12	12	62	4 No. 5
10	12	12	31	2 No. 5	12	12	51	3 No. 5	12	14	66	5 No. 5

A third design approach to consider for this footing is one in which the resistive moment of the passive soil pressure is included in the analysis. If we take this as the S value of 169 lb for the Group 4 soil, it offers a resisting moment of

$$M = (169)(6) = 1014 \text{ lb-in.}$$

which we subtract from the moment due to wind to obtain a new moment on the footing of

$$M = 3240 - 1014 = 2226 \text{ lb-in.}$$

With this new moment, combined with the dead load for the 18 in. wide footing we obtain a new eccentricity of

$$e = \frac{M}{N} = \frac{2226}{616}\ 3.61 \text{ in.}$$

which is only slightly in excess of the kern limit of 3 in. for the 18 in. wide footing.

If the footing is widened to 21 in., an analysis that includes the resistance due to passive soil pressure will produce an eccentricity approximately equal to that of the kern limit. We consider this to be a reasonable design solution for the footing.

Table 3.15 gives some recommended footings for freestanding walls for wind loads of 10, 15, and 20 lb/ft^2. Note that Section 2311(h) of the *Uniform Building Code* (Ref. 9) permits a one third reduction in the design wind pressure for fences not over 12 ft high. This reduction may not be permitted by other building codes, however. Footing widths given in the table have been determined on the basis of the procedure recommended in the design example, in which the eccentricity is limited to the kern limit, but the passive soil pressure is included in the analysis. The weight of the wall is approximately that obtained with concrete blocks of lighweight aggregate with voids partly filled with grout.

3.13 Footings for Shear Walls

When shear walls rest on bearing foundations the situation is usually one of the following:

1. The shear wall is part of a continuous wall and is supported by a foundation that extends beyond the shear wall ends.

2. The shear wall is a separate wall and is supported by its own foundation, in the manner of a freestanding tower.

We will consider the second of these two situations first. The basic problems to be solved in the design of such a foundation are the following:

Anchorage of the shear wall. The shear wall anchorage consists of the attachment of the shear wall to the foundation to resist the sliding and the overturning effects due to the lateral loads on the wall. This involves a considerable range of possible situations, depending on the construction of the wall and the magnitude of forces. We will not deal with this design problem in detail and refer the reader to the general discussions and numerous design examples given in *Simplified Building Design for Wind and Earthquake Forces* (Ref. 13).

Overturning effect. The overturning effect is taken into consideration by performing the usual analysis for the overturning moment due to the lateral loads and the determination of the safety factor resulting from the resistance offered by the dead loads and the passive soil pressure.

Horizontal sliding. Horizontal sliding is the direct, horizontal force resistance in opposition to the lateral loads. It may be developed by some combination of soil friction and passive soil pressure or may be transferred to other parts of the building structure.

Maximum soil pressure and its distribution. The magnitude and form of distribution of the vertical soil pressure on the foundation caused by the combination of vertical load and moment must be compared with the established design limits.

We will illustrate some of the issues involved in dealing with the last three of these problems in the two examples that follow.

Example 12 Independent Shear Wall Footing–Minor Load

The wall and proposed foundation are shown in Figure 3.69. The wall is assumed to function as a bearing wall as well as a shear wall, and the vertical loads applied to the top of the foundation are the sum of the wall weight and the support loads on the wall.

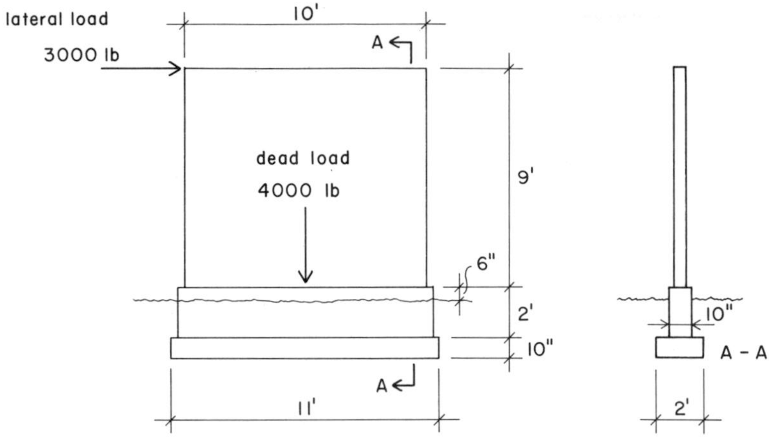

FIGURE 3.69. The shear wall and its loads–Example 12.

The following are design data and criteria:

Allowable soil pressure: 1500 lb/ft^2.

Soil type: Group 4, Table 29-B, *Uniform Building Code* (Ref. 9). See the appendix.

Concrete design strength: $f'_c = 2000$ psi.

Allowable tension on reinforcing: 20,000 psi.

The various forces acting on the foundation are shown in Figure 3.70. For the overturning analysis the usual procedure is to assume a rotation about the toe of the footing and to include only the gravity loads in determining the resistive moment. With these assumptions the analysis is as follows:

Overturning moment: $M = (3000)(11.83) = 35{,}490$ lb/ft.

Weight of foundation wall: $(2)(10/12)(10.5)(150) = 2625$ lb.

Weight of footing: $(10/12)(11)(2)(150) = 2750$ lb.

Weight of soil over footing: $(1.17)(1.5)(11)(80) = 1544$ lb.

Total vertical load: $4000 + 2625 + 2750 + 1544 = 10{,}919$ lb.

Resisting moment: $(10{,}919)(5.5) = 60{,}055$ lb-ft.

Safety factor: $SF = 60{,}055/35{,}490 = 1.69$.

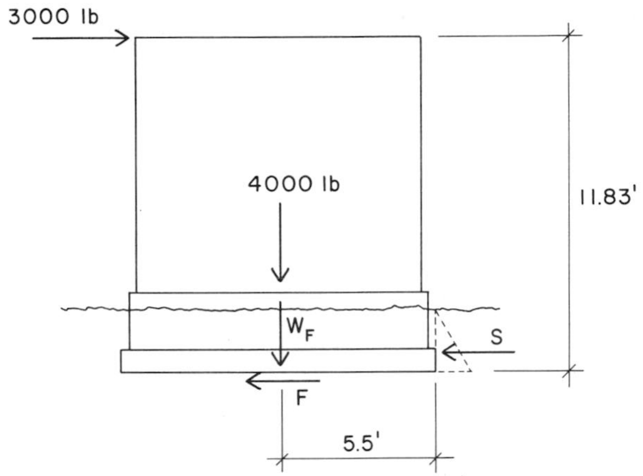

FIGURE 3.70. **Loading for the shear wall foundation analysis–Example 12.**

Since this safety factor is greater than the usual requirement of 1.5, the foundation is not critical for overturning effect.

For the soil group given the soil friction coefficient is 0.25, and the total sliding resistance offered by friction is thus

$$F = (0.25)(10{,}919) = 2730 \text{ lb}$$

Since we assume that the lateral load on the shear wall is due to either wind or seismic force, this resistance may be increased by one third. Thus, although there is some additional resistance developed by the passive soil pressure on the face of the foundation wall and the footing, it is not necessary to consider it in this example.

For the soil stress analysis we combine the overturning moment as calculated previously with the total vertical load to find the equivalent eccentricity as follows, deducting soil weight for N:

$$e = \frac{M}{N} = \frac{35{,}490}{9375} = 3.79 \text{ ft}$$

This eccentricity is considerably outside the kern limit for the 11 ft long footing ($\frac{11}{6}$, or 1.83 ft) so that the stress analysis must be done by the pressure wedge method, as discussed in Section 3.11. As illustrated in Figure 3.71, the analysis is as follows:

Distance of the eccentric load from the footing end is

$$5.5 - 3.79 = 1.71 \text{ ft}$$

Therefore,

$$x = (3)(1.71) = 5.13 \text{ ft}$$

$$p = \frac{2N}{wx} = \frac{(2)(9375)}{(2)(5.13)} = 1827 \text{ lb/ft}^2$$

Since this is less than the allowable design pressure with the permissible increase of one third $[p = (1.33)(1500) = 2000]$, the condition is not critical, as long as this type of soil pressure distribution is acceptable. This acceptance is a matter of judgment, based on concern for the rocking effect, as discussed in the example of the freestanding wall and illustrated in Figure 3.68. In this case, with the wall relatively short with respect to the footing length, we would judge the concern to be minor and would therefore accept the foundation as adequate.

The design considerations remaining for this example have to do with the structural adequacy of the foundation wall and footing. The short wall in this case is probably adequate without any vertical reinforcing, although it would be advisable to provide at least one vertical dowel at each end of the wall, extended with a hook into the footing. Compari-

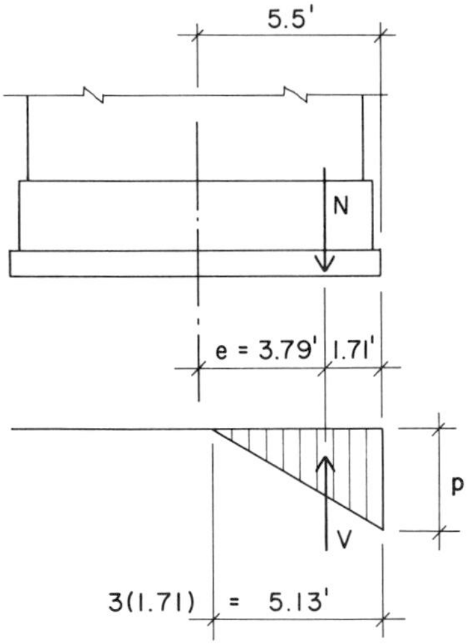

FIGURE 3.71. Vertical soil pressure—Example 12.

son with the entries in Table 3.1 indicate that the 2 ft wide footing is adequate without lateral reinforcing. Both the wall and footing, however, should be provided with some minimal longitudinal reinforcing for shrinkage and temperature stresses.

EXAMPLE 13 INDEPENDENT SHEAR WALL FOOTING—MAJOR LOAD

The wall and proposed foundation for this example are shown in Figure 3.72. Additional design data and criteria are as follows:

Allowable soil pressure: 3000 lb/ft^2.

Soil type: Group 4, Table 29-B, *Uniform Building Code* (Ref. 9). See the appendix.

Concrete design strength: f'_c = 3000 psi.

Allowable tension on reinforcing: 20,000 psi.

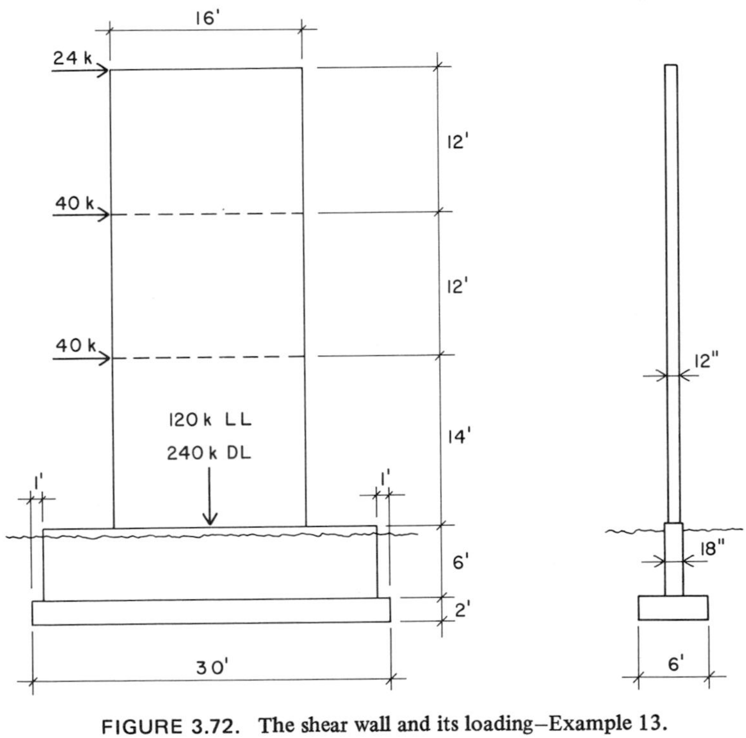

FIGURE 3.72. The shear wall and its loading–Example 13.

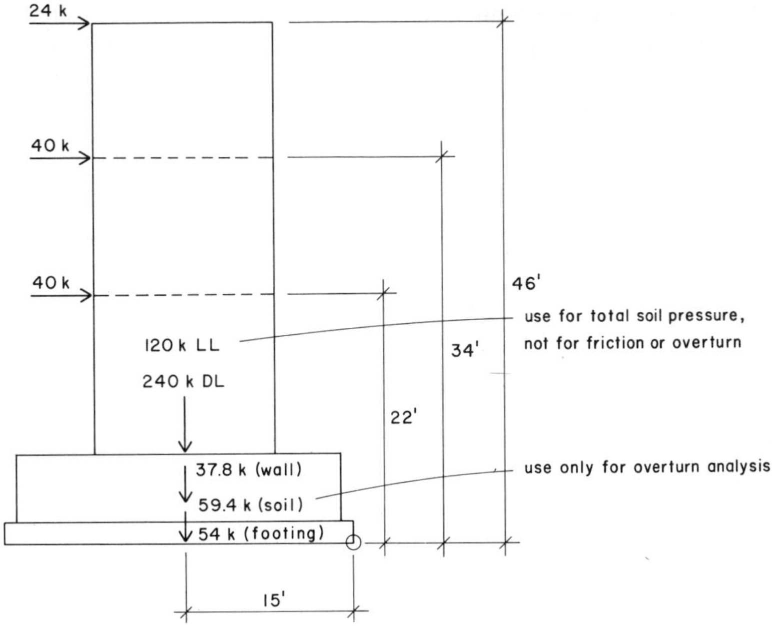

FIGURE 3.73. Loading for the shear wall foundation analysis–Example 13.

In this case the supporting foundation wall and footing are extended some distance past the end of the shear wall to increase the stability and reduce the soil pressures. The forces acting on the structure are shown in Figure 3.73. Following the usual procedure, we assume the overturning to be resisted only by the gravity forces and the rotation point for overturn to be at the toe of the footing. With these assumptions, the analysis is as follows:

Overturning moment:

$$M = (24)(46) + (40)(34) + (40)(22)$$
$$= 1104 + 1360 + 880 = 3344 \text{ k-ft}$$

Weight of foundation wall:

$$(1.5)(6)(28)(0.150) = 37.8 \text{ k}$$

Weight of footing:

$$(2)(6)(30)(0.150) = 54 \text{ k}$$

Weight of soil over footing:

$$(4.5)(5.5)(30)(0.08) = 59.4 \text{ k}$$

Total vertical load:

$$240 + 37.8 + 54 + 59.4 = 391.2 \text{ k}$$

Resisting moment:

$$M = (391.2)(15) = 5868 \text{ k-ft}$$

Safety factor:

$$SF = \frac{5868}{3344} = 1.75$$

Since this is greater than the required factor of 1.5, the foundation is not critical for the overturning effect.

For the soil group given, the soil friction coefficient is 0.25, and the total sliding resistance offered by friction is thus

$$F = (0.25)(391.2) = 97.8 \text{ k}$$

Since this is slightly less than the total horizontal load of 104 k, we will proceed with a determination of the additional resistance offered by the passive soil pressure on the end of the footing and foundation wall. Using the value for passive soil resistance for the Group 4 soil as given in Table 29-B of the *Uniform Building Code* (Ref. 9—see the appendix), the pressures are as shown in Figure 3.74 and are calculated as follows:

Table value for pressure/ft of depth: 150 lb/ft^2.

Maximum pressure at bottom of wall: (5.5)(150) = 825 lb/ft^2.

Pressure at bottom of footing: (7.5)(150) = 1125 lb/ft^2.

Total resistive forces:

On the end of the wall:

$$S_1 = \tfrac{1}{2}(1.5)(5.5)(0.825) = 3.4 \text{ k}$$

On the end of the footing:

$$S_2 = (2)(6)\ \frac{0.825 + 1.125}{2} = 11.7 \text{ k}$$

$$\text{total force} = S_1 + S_2 = 15.1 \text{ k}$$

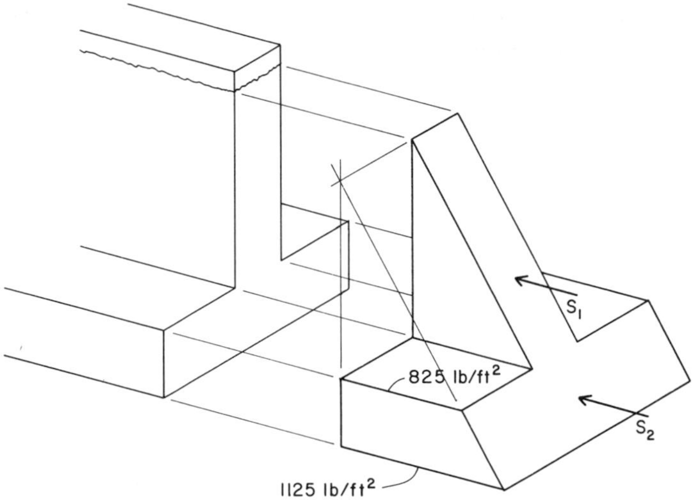

FIGURE 3.74. Horizontal soil pressures on the shear wall foundation–Example 13.

This increases the total resistive force due to the combination of sliding friction plus passive soil pressure to 112.9 k, which exceeds the total horizontal load.

For the vertical soil pressure on the bottom of the footing we consider the combined effect of the vertical load and the overturning moment. Although the passive soil pressure offers some resistance to the moment, it is relatively minor in this case and we will ignore it. The vertical load in this case should not include the weight of the soil over the footing, but must include the design live load. The loads and the resulting eccentricity are thus as follows:

Moment:

$$M = 3344 \text{ k-ft}$$

Vertical load:

$$N = 391.2 + 120 - 59.4 = 451.8 \text{ k}$$

Equivalent eccentricity:

$$e = \frac{M}{N} = \frac{3344}{451.8} = 7.40 \text{ ft}$$

This is considerably in excess of the kern limit of 5 ft for the footing and makes the design questionable. However, we will proceed with an analysis for the maximum soil pressure by the pressure wedge method as discussed in Section 3.11. Referring to Figure 3.75, the analysis is as follows.

The distance from the load to the edge of the footing is

$$15 - 7.4 = 7.6 \text{ ft}$$

Then

$$x = 3(7.6) = 22.8 \text{ ft}$$

$$p = \frac{2N}{wx} = \frac{(2)(451{,}800)}{(6)(22.8)} = 6605 \text{ lb/ft}^2$$

With the increase in allowable stress due to wind or seismic force, this would require a basic allowable soil pressure of

$$p = \tfrac{3}{4}(6605) = 4954 \text{ lb/ft}^2$$

which is greater than the given limit of 3000 lb/ft^2 in this example.

Reduction of the soil pressure requires an increase in the size of the footing. If this increase consists entirely of adding width, the gain is only a linear function of the increase. Increase in length is similar to

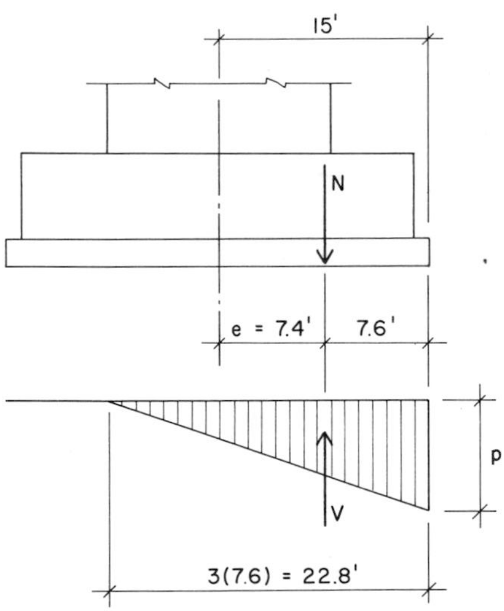

FIGURE 3.75. Vertical soil pressure—Example 13.

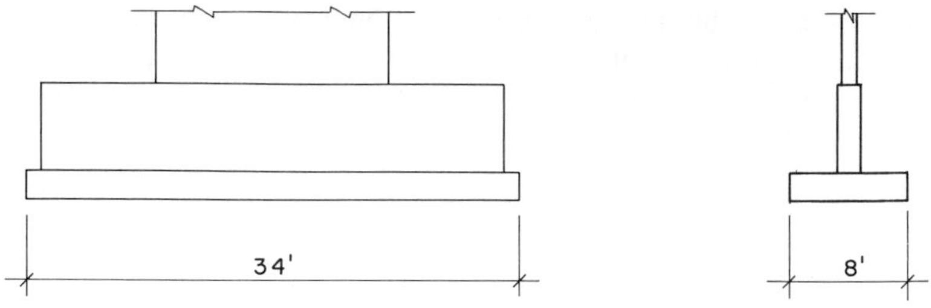

FIGURE 3.76. Modification of the foundation–Example 13.

adding depth to a beam section, which is considerably more effective in increasing bending resistance. However, in this situation increasing the footing length produces an increase in the cantilever distance for the foundation wall. We will therefore compromise with increases in both the width and length, as shown in Figure 3.76. These changes result in added weight of the foundation as follows:

New wall weight:

$$(1.5)(6)(32)(0.150) = 43.2 \text{ k}$$

New footing weight:

$$(2)(8)(34)(0.150) = 81.6 \text{ k}$$

New vertical load:

$$N = 360 + 43.2 + 81.6 = 484.8 \text{ k}$$

The new combined load analysis is thus as follows:

Eccentricity:

$$e = \frac{M}{N} = \frac{3344}{484.8} = 6.90 \text{ ft}$$

Distance from end of footing:

$$17 - 6.90 = 10.10 \text{ ft}$$

For the pressure wedge:

$$x = (3)(10.1) = 30.3 \text{ ft}$$

Maximum soil pressure:

$$p = \frac{2N}{wx} = \frac{(2)(484{,}800)}{(8)(30.3)} = 4000 \text{ lb/ft}^2$$

If the wedge type of soil stress distribution is acceptable, this is within the limit for the given soil with the permissible increase for wind and seismic loads. For this example the rocking phenomenon, as discussed for freestanding walls and illustrated in Figure 3.68, may be marginally critical. However, the overall height of the wall above the bottom of the footing is only 1.35 times the length of the footing, so the problem should only be a critical one if the soil is highly compressible or the building structure is highly sensitive to lateral deflections.

In this example we have assumed the shear wall and its foundation to be completely independent of the building structure, and have dealt with it as a freestanding tower. This is sometimes virtually the true situation, and the design approach that we have used is a valid one for such cases. However, various relations may occur between the shear wall structure and the rest of the building. One of these possibilities is shown in Figure 3.77. Here the shear wall and its foundation extend

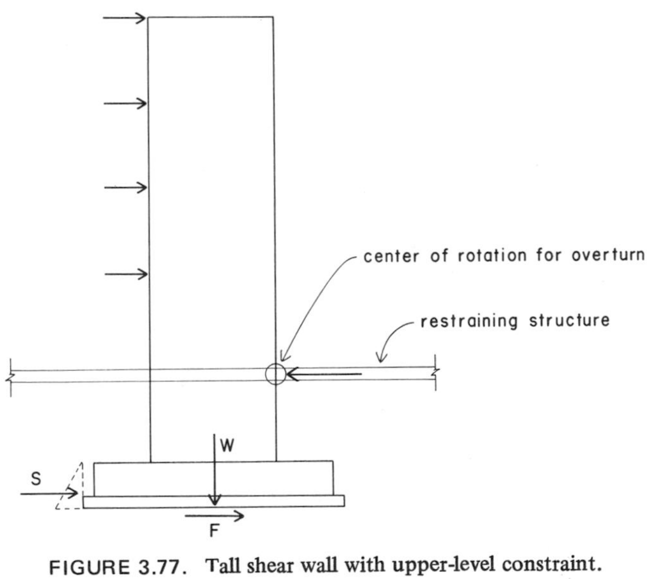

FIGURE 3.77. Tall shear wall with upper-level constraint.

some distance below a point at which the building structure offers a bracing force in terms of horizontal constraint to the shear wall. This situation may occur when there is a basement and the first floor structure is a heavy rigid concrete system. If the floor structure is capable of transferring the necessary horizontal force directly to the outside basement walls, the shear wall foundation may be relieved of the usual sliding resistance function.

As shown in Figure 3.77, when the upper-level constraint is present, the rotation point for overturn moves to this point. The forces that contribute to the resisting moment become the gravity load W, the sliding friction F, and the passive soil pressure S. The following example illustrates the analysis for such a structure.

Example 14 Shear Wall Footing—Upper Level Constraint

As shown in Figure 3.78, this structure is a modification of the one in the preceding example. We assume the construction to be the same as that shown in Figure 3.72, except for the added height of the wall and the constraint at the first-floor level. Data and criteria for design remain the same as in Example 13.

The only modification of the vertical loads from those determined for Example 13 is the additional basement wall. This added load is

$$w = (12\text{ ft})(16\text{ ft})(0.150\text{ k/ft}^2) = 28.8\text{ k}$$

Added to the total dead load calculated previously, this results in a new total deal load of

$$W = 391.2 + 28.8 = 420\text{ k}$$

The overturning analysis in this case begins with a comparison of the overturning moment and the resisting moment due to the dead load. If this does not result in the necessary safety factor of 1.5, we proceed to investigate the added forces that are necessary.

Overturning moment:

$$\begin{aligned} M &= (24)(38) + (40)(26) + (40)(14) \\ &= 912 + 1040 + 560 = 2512\text{ k-ft} \end{aligned}$$

Required dead load moment:

$$(2512)(SF\text{ of }1.5) = 3768\text{ k-ft}$$

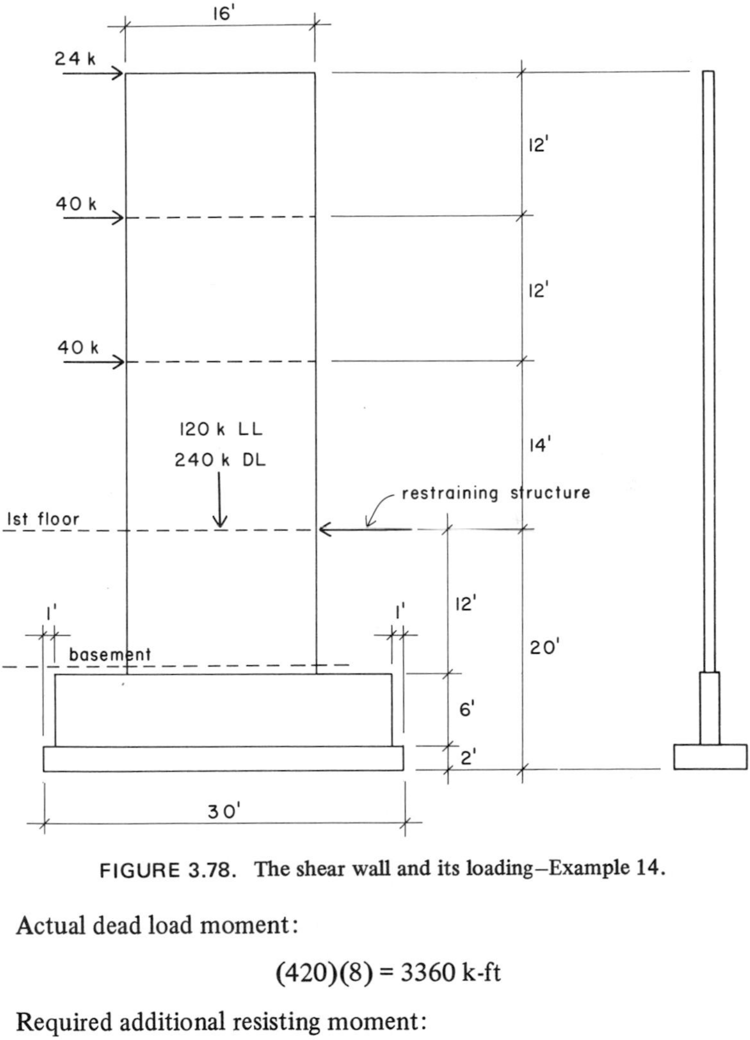

FIGURE 3.78. The shear wall and its loading–Example 14.

Actual dead load moment:

$$(420)(8) = 3360 \text{ k-ft}$$

Required additional resisting moment:

$$3768 - 3360 = 408 \text{ k-ft}$$

If we rely on the development of sliding friction for this moment, the necessary friction force is

$$F = \frac{408}{20} = 20.4 \text{ k}$$

which is quite a nominal force in view of the footing size and the magnitude of the dead load.

Since the friction is easily capable of the necessary added moment in this case, we do not need to consider the potential capability of added moment due to passive soil pressure. Were it necessary to do so, we would determine this potential force as was done for Example 13 and is illustrated in Figure 3.74.

Considering the equilibrium of the structure as shown with the forces in Figure 3.79, we can now determine the required force that must be developed by the constraining structure at the first-floor level. This will consist of the sum of the horizontal loads and the required friction force. Thus

$$R = H + F = 104 + 20.4 = 124.4 \text{ k}$$

If we consider the rotational stability of the wall to be maintained in the manner assumed in the preceding calculations, the vertical soil pres-

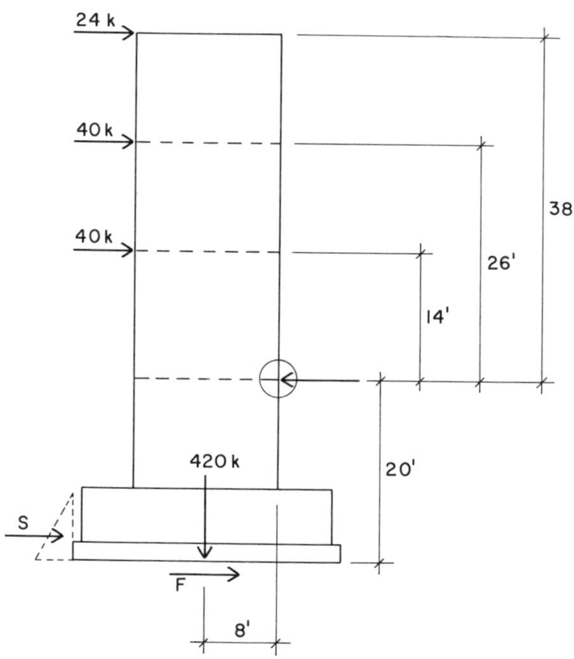

FIGURE 3.79. Loading for the shear wall foundation analysis–Example 14.

sure on the footing is relieved of any moment effect. Thus the pressure is simply that due to the vertical loads and is determined as follows:

Total vertical load:

$$420 \text{ k (dead load)} + 120 \text{ k (live load)} = 540 \text{ k}$$

Maximum soil pressure:

$$p = \frac{540}{(6)(30)} = 3 \text{ k/ft}^2$$

Since this is precisely the limit given, the footing is adequate in this example.

Another relationship that may occur between the shear wall structure and the rest of the building is that of some connection between the shear wall footing and other adjacent foundations. This occurs commonly in buildings designed for high seismic risk, since it is usually desirable to assure that the foundation system moves in unison during seismic shocks. This may be a useful relationship for the shear wall, in that additional horizontal resistance may be developed to add to that produced by the friction and passive soil pressure on the shear wall foundation itself. Thus, if the elements to which the shear wall foundation is tied do not have lateral load requirements, their potential friction and passive pressures may be enlisted to share the loads on the shear wall.

Shear walls on the building exterior often occur as individual wall segments, consisting of solid portions of the wall between openings or other discontinuities in the wall construction. In these situations the foundation often consists of a continuous wall and footing or a grade beam that extends along the entire wall. The effect of the overturning moment on such a foundation is shown in Figure 3.80. The loading tends to develop a shear force and moment in the foundation wall, both of which are one half of the forces in the wall. If the foundation wall is capable of developing this shear and bending, it functions as a distributing member, spreading the overturning effect along an extended length of the foundation.

The overturning effect just described must be added to other loadings on the wall for a complete investigation of the foundation wall and footing stresses. It is likely that the continuous foundation wall

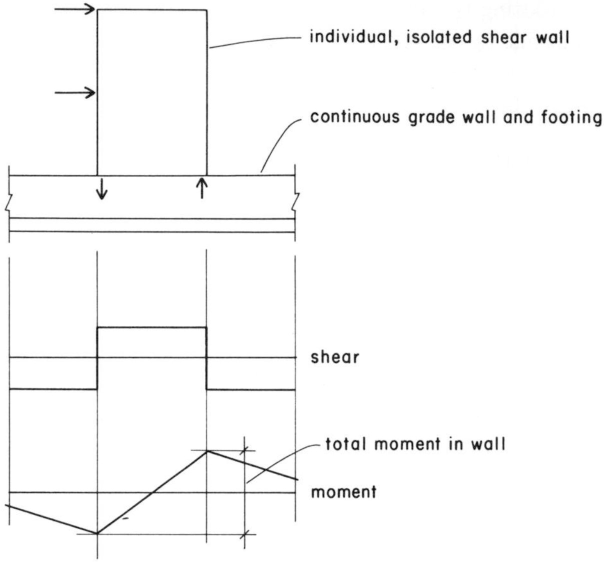

FIGURE 3.80. Isolated shear wall on a continuous foundation.

also functions as a distributing member for the gravity loads, as discussed in Section 3.10.

3.14 Tower Footings

The design problems for tower footings are essentially the same as those discussed for freestanding walls and isolated shear walls. The tower footing must perform adequately in terms of the three basic considerations: overturning moment, horizontal sliding, and vertical soil pressure. As the height of the tower increases, the predominant considerations tend to become the overturning moment and the movement of the structure associated with it.

The horizontal deflection at the top of a vertically cantilevered structure, such as that shown in Figure 3.81, is produced by a number of things. These include:

The flexing of the tower structure itself.

Yield in the tower-to-base connection.

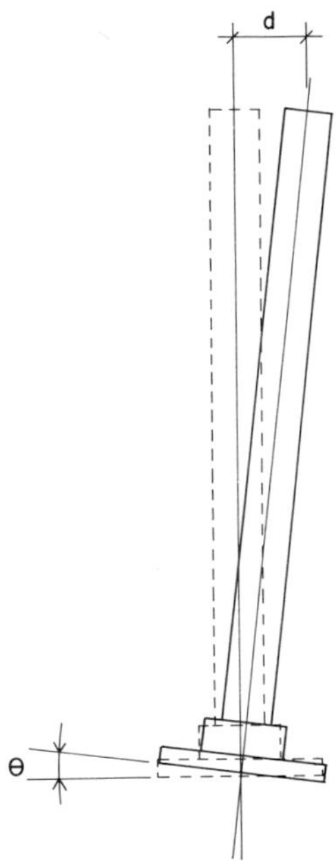

FIGURE 3.81. Lateral movement of a tower structure.

Structural deformations in the tower base.

Tilt of the base due to the uneven soil stresses caused by the combined vertical load and moment.

If the tower foundation is a bearing footing, some rotation of the base is inevitable, unless the footing is placed on solid rock. As the tower height increases this rotation can become quite critical. For example, with a 300 ft tall tower, a rotation of only 0.5 degree at the base will produce a horizontal movement of approximately 30 in. at the top of the tower.

The design of foundations for large towers is beyond the scope of this book. For modest structures the design may be performed in the manner illustrated for the freestanding wall and the isolated shear wall. If the horizontal deflection of the tower is considered to be critical, it is advisable to keep the equivalent eccentricity well within the kern limit of the footing. And when the tower height exceeds 50 ft or so, it is highly questionable to use a bearing footing when the soil is a soft clay, a loose sand, or some other material subject to considerable deformation.

3.15 Miscellaneous Problems of Shallow Bearing Foundations

A number of problems occur in the general design of foundation systems that utilize shallow bearing footings. While the individual problems of various types of foundation elements are discussed in various other parts of this book, the following are some problems that are often shared by the several elements that constitute the complete foundation system for a building.

Equalizing of Settlements. It is usually desired that all of the elements of a foundation system settle the same amount. If part of a wall settles more than another, or if columns settle more than walls, there are a number of problems that can result, such as:

Cracking of walls, especially those consisting of rigid materials such as masonry, plaster, and concrete.

Jamming of doors and operable windows.

Fracturing of plumbing or electrical conduits incorporated into the structure.

Producing of undesirable stress or stability conditions in structures that have some degree of continuity, such as multispan beams or rigid frames of steel or reinforced concrete.

A technique sometimes used to reduce these problems is to design for so-called equalized settlements. This is a process in which the sizes of bearing elements are determined on the basis of developing equal vertical settlement, rather than producing a common maximum soil pressure. The ease of accomplishing this depends on several factors, including the specific soil conditions. The simplest case occurs when all the footings are at the same approximate level and all bear on the same type of soil. In this case the technique most often used is to design for a relatively constant pressure under the dead load, since this most often represents the critical loading condition for settlements. The process is as follows:

1. Design loads are established for each bearing element, separating the live load from the dead load.
2. The element with the highest ratio of live load to dead load is

selected and designed for a total soil pressure using the limit of allowable bearing pressure.

3. The pressure under the dead load only is determined for the selected control element. This represents the desired constant pressure for design.
4. The required plan dimensions for all other bearing elements are determined, using only their dead loads and the control pressure determined in Step 3.
5. With their plan sizes established, the other elements are then designed structurally for their own individual design soil pressures under the total loading of the dead and live loads.

The following example illustrates the use of this design method in its simplest form.

EXAMPLE 15 DESIGN FOR EQUALIZED SETTLEMENT UNDER DEAD LOAD

The building plan shown in Figure 3.82 indicates a number of wall and column elements. It is desired that the footings provided for these structural elements produce soil pressures that are equal under the dead load only and do not exceed 2000 lb/ft^2 under the total load. The loads for the individual elements are given in Table 3.16, together with the determination of the live-to-dead load ratio for each element. The dead loads shown include an estimate of the footing weight.

Since the footing for Column A has the highest ratio of live-to-dead load, it is selected for determination of the design dead load pressure.

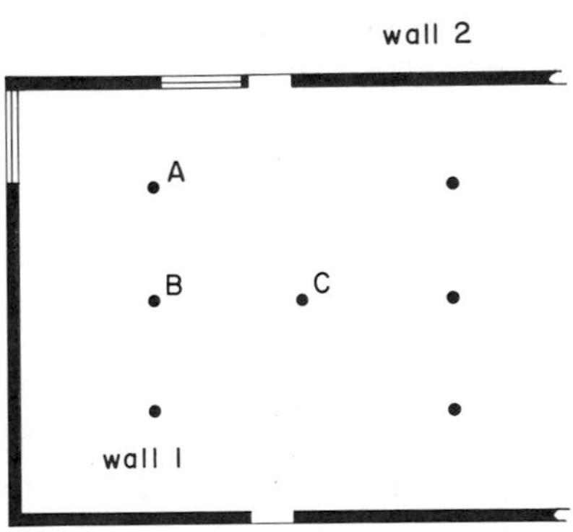

FIGURE 3.82. Plan of the building—Example 15.

TABLE 3.16. Foundation Loads (Example 15)

Load for:	Live Load (k)	Dead Load[a] (k)	Live Load/Dead Load (ratio)
Column A	24	12	2.00
Column B	30	20	1.50
Column C	70	50	1.40
Wall 1	1.2/ft	2.0/ft	0.60
Wall 2	1.8/ft	3.0/ft	0.60

[a] Includes estimated footing weight.

This is the only footing that will be designed in the usual manner, using the maximum allowable soil pressure to determine its required plan dimensions. For Footing A:

Required area of footing:

$$A = \frac{36{,}000}{2000} = 18 \text{ ft}^2$$

Select: 4 ft 3 in. square, 12 in. thick (from Table 3.5, or by actual calculation).

Dead load pressure:

$$p = \frac{12{,}000}{(4.25)^2} = 664 \text{ lb/ft}^2$$

This pressure is now used to determine the required plan size of the other footings. With their plan dimensions thus established, the net design pressure under the total load (dead load plus live load omitting the footing weight) is determined for each footing and is used to perform the structural design of the footing. This process is illustrated in Table 3.17.

The effectiveness of the technique illustrated in this example is limited. Some of the conditions that may require a different design approach are the following:

Nonuniform soil conditions. Where the soil conditions vary for separate footings, maintaining a constant soil pressure may not assure

TABLE 3.17. Footing Design Data (Example 15)

Footing for:	Dead Load (k)	Required Size (ft)	Selected Size	Total Load (k)	Total Soil[a] Pressure (k/ft^2)	Net Design[b] Soil Pressure (k/ft^2)
Column A	12	$(4.25)^2$	4.25 ft^2 by 12 in. thick	36	1.99	1.840
Column B	20	$(5.49)^2$	5.50 ft^2 by 12 in. thick	50	1.65	1.500
Column C	50	$(8.68)^2$	8.67 ft^2 by 18 in. thick	120	1.60	1.375
Wall 1	1.2/ft	1.81	1.75 ft wide by 10 in. thick	3.2/ft	1.83	1.710
Wall 2	1.8/ft	2.71	2.67 ft wide by 10 in. thick	4.8/ft	1.80	1.680

[a] Including weight of footing.
[b] For stress design of footing; weight of footing deducted.

equal settlements. Nonhorizontal soil strata, pockets of poor soil, or weak underlying strata affected by large footings are some situations of this type. These conditions may require actual settlement calculations for each individual footing.

Time-dependent settlements. Granular soils, consisting of clean sand and gravel, tend to settle instantly under the application of the maximum load. In this case the live load may have more effect than is given credit in the process illustrated in Example 15. Soft clays, on the other hand, tend to produce long-time, continuous settlements, for which the dead load is more truly critical.

Significant range in load magnitudes. In some buildings, parts of the foundation system may be quite lightly loaded, while others sustain very heavy loads. An example of this is a building with adjacent high-rise and low-rise portions. This situation often makes the equalizing of pressures unfeasible. If significant differences in settlements are involved, it may be necessary to provide some structural separation between the adjacent parts of the building. If settlements can be reasonably accurately predicted, it may be possible to provide for independent vertical settlements during the building construction, with the adjacent parts arriving at some alignment when construction is complete.

Proximity of Foundation Elements. Building planning often results in situations in which separate parts of the building structure are located so that their foundations are close together. Some of the situations of this type and the design problems that result are the following:

Closely spaced columns. Columns are occasionally so closely spaced that it is not possible to use the usual square column footing under each column. When this occurs, the three main options are to place an oblong footing under one column with one dimension restricted to that required to clear the adjacent column, to place oblong footings under both columns, or to use a single footing for both columns. Oblong footings are discussed in Section 3.7 and combined footings are discussed in Section 3.8.

Closely spaced buildings. When new construction must be placed very close to an existing building, the excavation and foundation construction problems must be studied very closely. Excavation must be performed in a manner that does not cause settlement or collapse of the adjacent, existing construction. If the new footings must be lower than the existing ones, it will most likely be necessary to underpin the adjacent building. This is a major engineering problem and should be undertaken only by persons with considerable training and experience in foundation design. If the two adjacent constructions must be connected, a critical foundation design problem is that of equalizing the settlement of the two adjacent parts, which may be complicated by the fact that the soil stress due to the new construction may cause some additional settlement of the existing structure. When the second construction is anticipated at the time of the original work, it is sometimes possible to provide a combined footing that will eventually support both of the adjacent structures, as shown in Figure 3.83. This will assure alignment of the adjacent constructions, but may present another problem: rotation of the footing under the highly eccentric load due to the original construction. Another technique is to cantilever the structure of one or both of the adjacent constructions so their footings may be spread a sufficient distance to eliminate the problems of undermining or soil stress overlap.

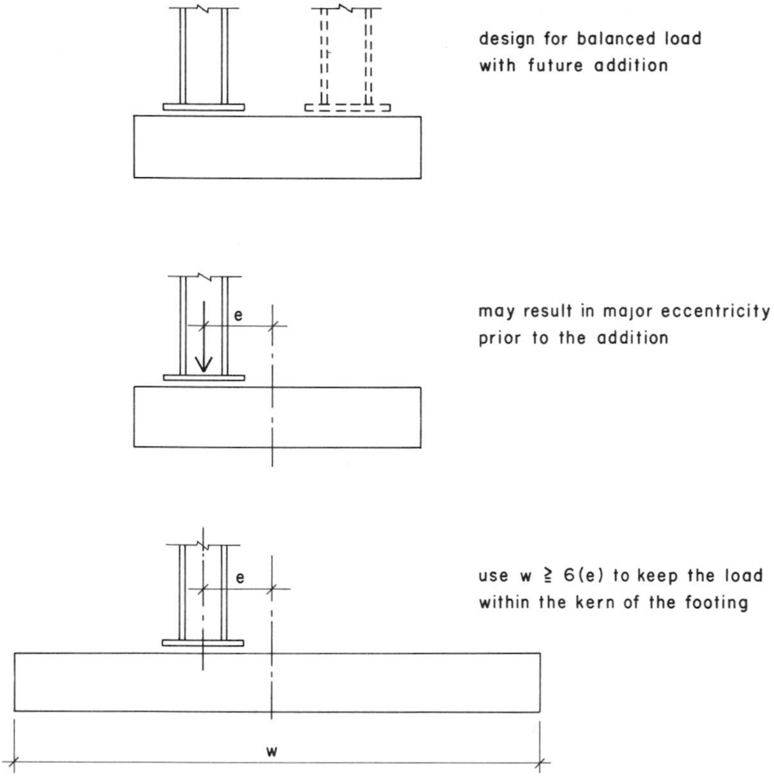
design for balanced load
with future addition
e
may result in major eccentricity
prior to the addition
e
use w ≧ 6(e) to keep the load
within the kern of the footing
w

FIGURE 3.83. Combined footing for a future addition.

Footings adjacent to pits, tunnels, etc. Columns or bearing walls must sometimes be located near steam tunnels, elevator pits, underground utility vaults, or other structures that must be placed at an elevation near or below the normal elevation for the footings. When loads on the walls or columns are relatively light, it may be possible to combine the structures and their foundations. When the structural loads are high, however, it is usually necessary to drop the footings for the walls or columns below the adjacent construction. If this elevation change is considerable, it may create a problem with regard to other footings in the vicinity. This is treated in the discussion that follows.

Adjacent Footings at Different Elevations. When individual footings are horizontally closely spaced but occur at different elevations, a number of potential problems may be created. As shown in Figure 3.84, some of these are as follows:

Disturbance of the upper footing. Excavation for the lower footing may result in settlement or collapse of the upper footing, if the upper footing is already in place and carrying a significant load. Even when the two excavations are performed at the same time, however, the

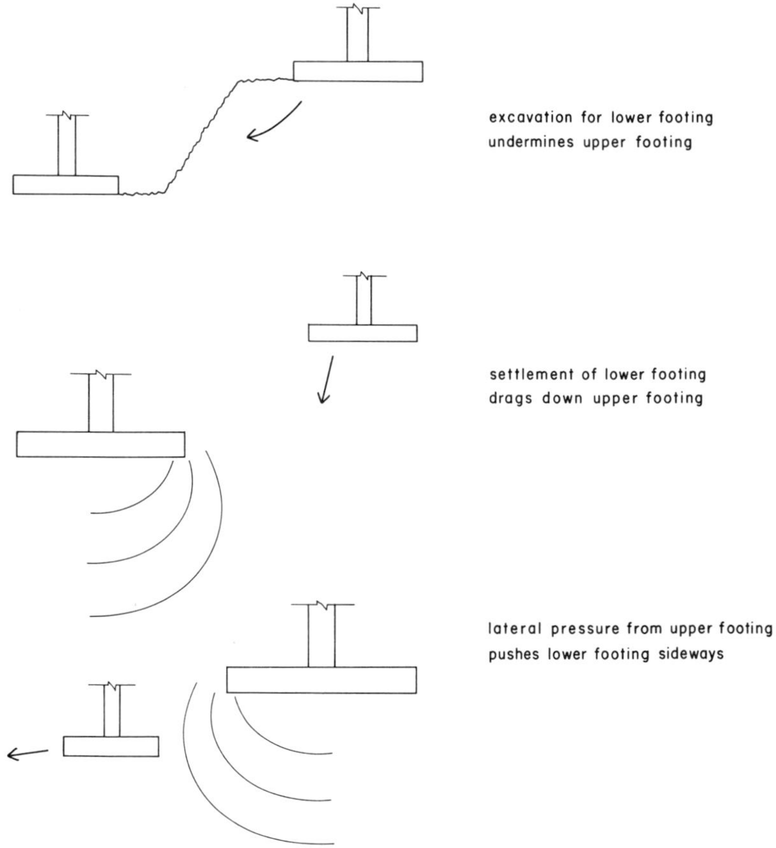
excavation for lower footing
undermines upper footing
settlement of lower footing
drags down upper footing
lateral pressure from upper footing
pushes lower footing sideways

FIGURE 3.84. Problems with adjacent footings at different elevations.

deep cut required for the lower footing may significantly disturb the soil under the upper footing. There are a number of variables in this situation, so that each case must be carefully studied. The relative size and load on the two footings must be considered. Also of major concern is the effect of a steep cut on the soil. In highly cohesionless material, such as clean sand or gravel, this will surely cause considerable disturbance of the soil structure. In a soft clay there will be an oozing due to the pressure relief.

Additional settlement of the upper footing. If the lower footing is large, its pressure may spread sufficiently through the soil mass to cause some additional settlement of the upper footing, as it is dragged down with the lower one.

Lateral pressure effect on the lower footing. If the upper footing is large, the horizontal spread of pressure beneath it may produce some lateral movement of the lower footing.

The critical design limit in these situations is the relation of the vertical separation to the horizontal distance between the footings, as shown in Figure 3.85. This limit is very much a matter of judgment and cannot be generalized for all situations. We hesitate to recommend any single number for this ratio, since it would be highly conservative for some situations and potentially critical for others. Some building codes or building regulatory agencies set limits for this ratio, although sometimes it is possible to make a case for an exception where good information about soil conditions is available and a careful engineering analysis can be performed.

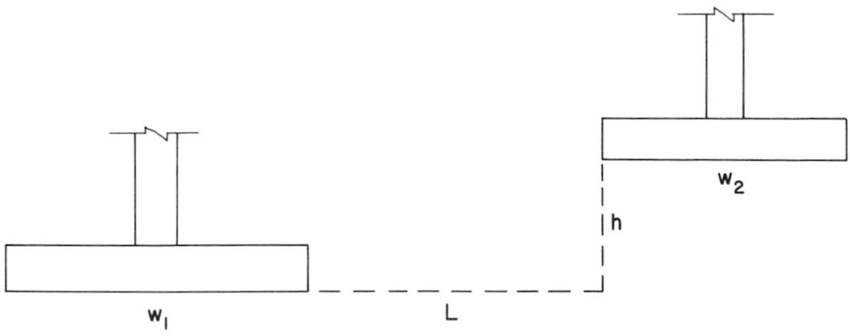

FIGURE 3.85. Dimensional relationships for footings at different elevations.

Although the ratio of the dimensions h and L is a critical concern with regard to the various effects illustrated in Figure 3.84, of equal concern is the actual value of L. When this distance is close to or less than the dimension of the larger footing, the soil stress may cause difficulties, even though the footings are at the same elevation. Thus a value of h/L of one half or less does not mean the design is conservative. Conversely, with the same soil conditions, when L is several times the dimension of the larger footing, a considerable elevation difference may be tolerated.

Another problem involving differences in footing elevations occurs when the bottom of a continuous foundation wall must be lowered at some point. A common situation of this type occurs when a building has only a partial basement, requiring the foundation wall to be considerably lower in the area of the basement. As shown in Figure 3.86, the solution to this problem is to either slope or step the wall footing between the two levels. Unless the angle of the required slope is quite low (h/L of $\frac{1}{5}$ or less) the usual preference is for a stepped footing.

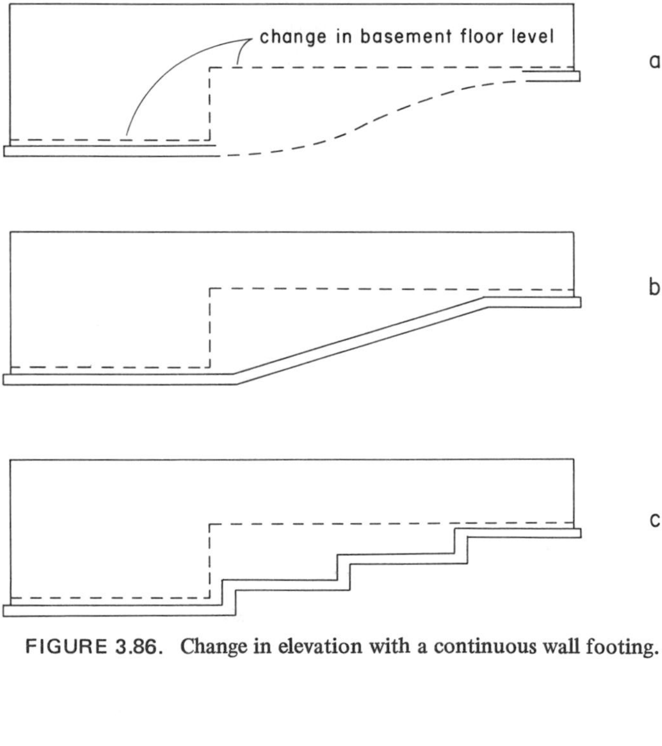

FIGURE 3.86. Change in elevation with a continuous wall footing.

There are three critical considerations for the design of the stepped footing.

1. ***The length of the step.*** If the step length is too short, the individual steps will have questionable validity as individual footings, and the footing is effectively the same as a sloped one. As shown in Figure 3.87, the toe portion of the step is essentially unusable for bearing. Thus if the step is very short, the length remaining for use as a bearing footing may be quite minor.
2. ***The height of the step.*** The higher the step, the longer the unusable toe portion of the flat step. This has something to do with the soil type and is related to excavation problems. In soils that are unfeasible to excavate with a vertical cut it may be necessary to slope the stepped cut, as shown in the lower part of Figure 3.87.

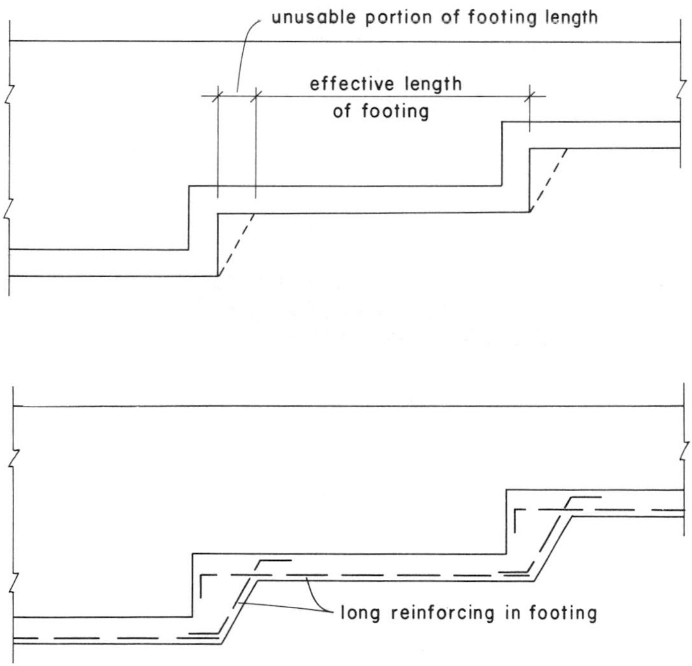

FIGURE 3.87. Considerations for stepped wall footings.

3. ***The angle of the step.*** The step angle–the h/L ratio as shown in the figures–is essentially the same problem as was discussed for the individual footings and illustrated in Figure 3.85.

As in the case of individual footings, the details of stepped footings may be regulated by building codes or regulatory agencies. For a generally conservative design, and in the absence of other design limitations, we recommend the following.

Limit the step length to not less than three times the footing width (footing dimension perpendicular to the wall plane).

Limit the step height to not more than $1\frac{1}{2}$ times the footing width, or 2 ft, whichever is smaller.

Limit the h/L ratio to one third or less.

Footings on Backfill. In general it is desirable that bearing footings be placed on undisturbed soil, undisturbed meaning that the soil has not been previously excavated. If this is not feasible, the choices are limited to the use of a deep foundation or footings placed on fill materials. In some situations, when footing loads are relatively light, it may be reasonable to consider the latter option. Some such situations are the following:

1. When a relatively thin layer of undesirable material is encountered at the level desired for the footings. In this event it may be possible to deposit a thin layer of highly compacted material to replace the poor soil, and to end up with a much improved bearing condition.
2. When the soil at the desired bearing level is highly sensitive to disturbance by the excavation and foundation construction activites. This may be dealt with in a manner similar to the previous situation. If the footings are small, the pressures developed below the level of the fill may be able to be sustained by the weaker materials.
3. When the excavation of some unanticipated buried object, such as a large tree root, old sewer line, or large boulder, leaves the usable level for bearing a single footing considerably below that of adjacent footings. As in the previous cases, if a small amount of fill can be used, it may be a better choice than dropping all the adjacent footings.

For any of these situations, or other ones involving placing footings on backfill, expert advice should be sought regarding the feasibility of the decision, the type of material to be used for the backfill, and the construction process to be used to obtain the necessary degree of compaction of the fill.

Study Aids

Words and Terms. Using the glossary and the text of the preceding chapter for reference, review the meaning of the following words and terms.

Allowable bearing stress.
Anchorage.
Combined footing.
Dead load.
Equalized settlement.
Footing.
Freestanding wall.
Friction.
Kern.
Live load.
Mat foundation.
Overturning effect.
Restoring moment.
Shallow bearing foundation.
Shoring.

Questions

1. What are the various potential functional requirements for foundation walls?
2. What is the usual basis for establishing a minimum width for a wall footing?
3. Why is the square, rather than round, form commonly used for a column footing?

4. What are the various possible functional uses for piers in bearing foundation systems?
5. Why is it considered desirable to keep the resultant of the forces acting on a moment-resistive footing within the kern limit of the footing?

Problems

See selected answers following the appendix.

1. Select an entry in one of the tables of footing designs (Tables 3.1, 3.2, 3.4 and 3.5) and verify the table data by performing the calculations as illustrated in the example problems in the text.
2. Using strength design methods and the latest *ACI Code* requirements, redesign one of the entries in Table 3.2 or 3.5. Use f_y = 50,000 psi and assume the total load to be 50% live load and 50% dead load.
   a. Use the footing dimensions and reinforcing given and determine a new allowable load.
   b. Use the table load entry and determine any possible savings in concrete and steel.
3. Select a column footing of reasonably large size from Table 3.5, assume that the width in one direction is limited to $\frac{3}{4}$ that given in the table, and design the required oblong rectangular footing. Compare the steel and concrete quantities with those in the table entry.
4. Design a single footing to carry two columns with the loads and center-to-center column spacings given. Use f_c' = 3000 psi, f_s = 20,000 psi, maximum allowable soil pressure = 3000 lb/ft².

	Column 1 Load (k)	Column 2 Load (k)	Spacing (ft)
A	100	100	6
B	200	200	10
C	100	200	8

5. Design a cantilever footing for the data given. The edge of the footing is limited to 12 in. from the center of Column 1 in one direction. Use $f_c' = 3000$ psi, $f_s = 20{,}000$ psi, maximum allowable soil pressure = 3000 lb/ft^2.

	Column 1 Load (k)	Column 2 Load (k)	Spacing (ft)
A	100	200	10
B	200	400	15
C	300	400	18

6. Design a footing for a freestanding wall with the following data. Design for vertical pressure so that the resultant of the forces stays within the kern limit of the footing. Place the bottom of the footing at 18 in. below grade. Use $f_c' = 3000$ psi, $f_s = 20{,}000$ psi, passive soil resistance of 150 lb/ft^2 per ft of depth, and a friction coefficient of 0.25.

	Height of Wall Above Grade (ft)	Weight of Wall (lb/ft^2)	Allowable Soil Pressure (lb/ft^2)	Wind Load (lb/ft^2)
A	4	55	1000	10
B	5	65	1500	10
C	8	90	2000	20

7. Using Tables 3.2 and 3.5, select approximate sizes for the footings for the walls and columns for the following buildings. Design for a condition of equalized pressure under dead load and a maximum pressure of 2000 lb/ft^2.

	Wall A Load (k/ft)		Wall B Load (k/ft)		Column C Load (k)		Column D Load (k)	
Building	*DL*	*LL*	*DL*	*LL*	*DL*	*LL*	*DL*	*LL*
A	1.2	0.6	1.8	3.1	26	48	47	72
B	0.6	1.0	1.3	2.1	18	30	25	42
C	3.1	1.8	4.6	2.4	30	50	60	80

4

Design of Deep Foundations

In most instances deep foundations are utilized only where it is not possible to have shallow foundations. The decision to use deep foundations, the selection of the type of system, and the design of the elements of the system should all be done by persons with considerable knowledge and experience in foundation design. The material in this chapter is provided to help the reader gain familiarity with the types of systems, their capabilities and limitations, and some of the problems involved in utilizing such systems for building foundations. This familiarity is important for all building designers, even though they may need expert advice and assistance for completion of the foundation design.

4.1 Establishing the Need for Deep Foundations

The most common reasons for using deep foundations are the following:

1. ***Lack of adequate soil conditions for bearing footings.*** As shown in Figure 4.1, there are a number of soil conditions that may make it impossible to place the usual bearing foundation elements near the bottom of the building. The deep foundation thus becomes essentially a means of reaching a desirable bearing level at some distance from the bottom of the building.

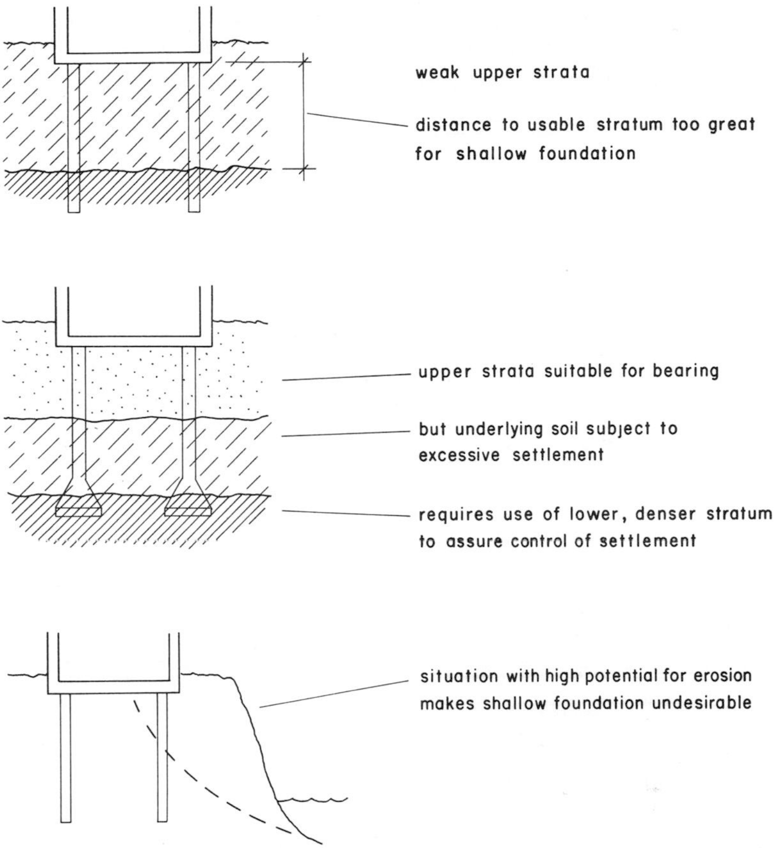
weak upper strata
distance to usable stratum too great
for shallow foundation
upper strata suitable for bearing
but underlying soil subject to
excessive settlement
requires use of lower, denser stratum
to assure control of settlement
situation with high potential for erosion
makes shallow foundation undesirable

FIGURE 4.1. Situations requiring deep foundations.

2. ***Heavy loads on the foundations.*** In some cases the soil at upper levels may be sufficient for the use of bearing elements for relatively light loads, but the size of footings required, the need for limited settlement, or other factors occurring with excessively heavy loads, may require a deep foundation. High-rise buildings, long span structures, and construction of massive elements of concrete or masonry are cases in which loads may become considerable and may exceed the simple bearing capabilities of ordinary soils.

3. ***Potential instability of ground-level soil.*** Use of deep foundations may be necessary where soil at the level of the bottom of the building is subject to erosion, subsidence, slippage, decomposition, or other forms of change in the soil structure or the general state of the soil mass. This situation sometimes occurs at waterfront and hillside locations. In these cases the use of deep foundations is a means for anchoring the building to a more reliable, stable ground mass.
4. ***Support of structures highly sensitive to settlement.*** In some situations settlement of foundations is highly critical. Examples are buildings with stiff rigid frame structures and buildings housing equipment that requires precise and continuous alignment. Bearing foundations will settle on almost any soil other than solid bedrock. Deep foundations tend to have little settlement, especially when the elements are carried down to rock or to a highly consolidated soil stratum.

As stated previously, deep foundations are usually used only in situations in which shallow footings are not possible. The primary reason is cost. Where ordinary footings can be used and their size is not excessive, the cost of deep foundations will seldom be competitive. When footings cannot be used, this cost must be borne. For marginal situations—where footings are possible but their size is excessive—it may be necessary to perform a cost analysis of alternatives. Such an analysis will usually require relatively complete designs of the alternate systems and can be further complicated when the alternate foundation systems require significant differences in the building structure.

Foundation design and construction practices are often strongly influenced by the regional location of a building. This may be partly because of natural phenomena such as the local climate and soil conditions. Where there is some history of construction in the area, however, it may also be due to local experience with particular foundation systems or construction procedures. In the case of deep foundations, an important consideration is often the availability of local contractors with the necessary equipment and expertise for particular types of construction.

Most deep foundation elements—piles or piers—are installed by special foundation contractors, often using particular types of equipment or construction elements. Thus while the design of a foundation may be developed with a basic type of deep foundation in mind, the final design and detailing may have to be delayed until a particular contractor is selected.

In areas of considerable construction activity, especially near to large metropolitan areas, there may be competition between several companies specializing in deep foundations. In more remote areas there may be only a few companies, or even a single company, available for such work. Since the equipment necessary for this work is often quite large and not easily transportable, the distance from a contractor's home base to the building site is a critical cost factor.

Before beginning the design of any building foundation it is advisable to learn as much as possible about local experiences and practices. Evolution of technology and advances in design techniques may result in more intelligent designs, but it is still wise to proceed from a base of knowledge of previous construction experience. The proven need for and general feasibility of deep foundation construction is an area particularly sensitive to these considerations.

4.2 Types of Deep Foundations

As shown in Figure 4.2, the common basic types of deep foundations are the following:

Friction piles. These usually consist of timber, steel or precast concrete shafts that are forcibly inserted into the ground, most often by

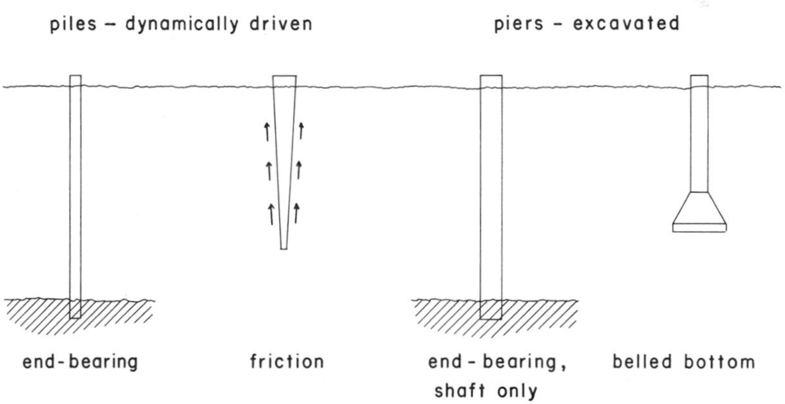

FIGURE 4.2. Types of deep foundations.

dynamic driving similar to a hammer pounding a nail into a piece of wood. Vertical load-carrying capability is developed by surface friction between the pile and the ground. Specific rating of load capacity is commonly established on the basis of the measured effort required for advancing the pile the last few feet of penetration.

End-bearing piles. These are usually elements similar to those used for friction piles, although in this case they are driven a specific distance in order to lodge their ends in some highly resistive soil stratum or in rock. While considerable skin friction may be developed during the advancement of the pile, the major load capacity is developed at the point of the pile. The chief function of the upper soil strata in this case is to maintain the lateral stability of the long compression member.

Piers or caissons. For various reasons it may be better to place an end-bearing element by excavating the soil down to the level at which bearing is desired and then backfilling the excavation with concrete. The structural function of such an element is essentially the same as that of an end-bearing pile. An advantage of this technique is that the material encountered at the bottom of the shaft can be inspected before pouring the concrete. The capacity of many end-bearing piles, on the other hand, can only be inferred on the basis of the length driven, the difficulty of driving, or by load testing the driven pile.

Belled piers. When piers bear on solid rock they usually have an end-bearing capacity approximately equal to that of the concrete shaft in column action. When they bear on soil, however, they must usually be enlarged at their ends in order to increase the bearing area. The usual conical form of this end enlargement yields the term "belled" to describe such an element.

Pile foundations are discussed in Section 4.3 and piers in Section 4.4. The following are some of the problems encountered with both types of elements and some of their relative merits.

Piles must generally be driven in groups. One reason for this is their limited individual capacities, another is the problem of precisely controlling their locations during the placing process. Even when loads are small, it is generally not feasible to place a concentrated column load on top of a single pile since precise alignment of the column and pile is not feasible. The preferred minimum group for a concentrated load is three piles, typically arranged in a symmetrical, triangular pattern, as shown in Figure 4.3. With piles spread a minimum distance, usually at least 30 to 36 in. center to center, a small mislocation of the column and pile group centroids can usually be tolerated. If one of the piles is slightly mislocated during driving it may be possible to alter the location of the rest of the group to compensate for this, as shown in the illustration. However, if the mislocation cannot be adjusted for in this manner, it becomes necessary to add more piles to the group in order to regain a reasonable alignment of the column and the group centroid.

When piles are driven in tight clusters and closely spaced groups, the driving of each successive pile often causes lateral movement of the top of previously driven piles. Thus it is necessary to check the final location of all piles before proceeding to the completion of the foundation construction. If substantial movements have occurred it may be necessary to add piles to some groups to once again align the group centroids with the required column locations.

In order to achieve the transfer of load from a column to the several piles in a group it is necessary to use a stiff reinforced concrete cap on top of the piles. This element functions like an isolated column footing and also accommodates anchor bolts or reinforcing dowels.

Driving piles usually requires quite heavy equipment. If the building site is remote or inaccessible, the problem of moving the equipment to the site and into position for driving can be a major concern. Pile driving is also noisy and jars the ground, which can be a problem for the neighbors.

Piers can be produced in a large range of sizes. Small piers of 12 to 36 in. diameter can be drilled with an auger-type rig, similar to the simple truck-mounted equipment used for well drilling or post-hole drilling. This equipment is usually relatively light and portable compared to that required for pile driving. Belled ends can be excavated with the drilling rig using a spreader device once the pier shaft has been drilled.

For both piles and drilled piers limiting soil conditions are required. The presence of many large boulders may preclude either operation. Loose sand or a high water table may make it difficult to maintain the excavation while placing the concrete. Wet clay may offer resistance to pile driving or prove impossible to drill out so that excavation may have to be hand advanced. The limiting conditions for each type of

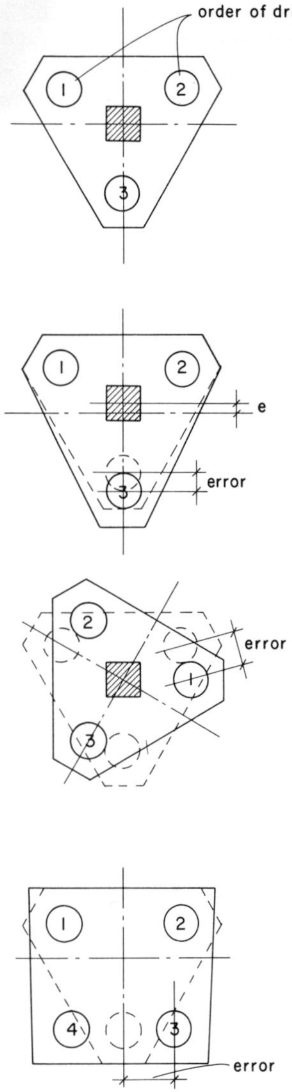

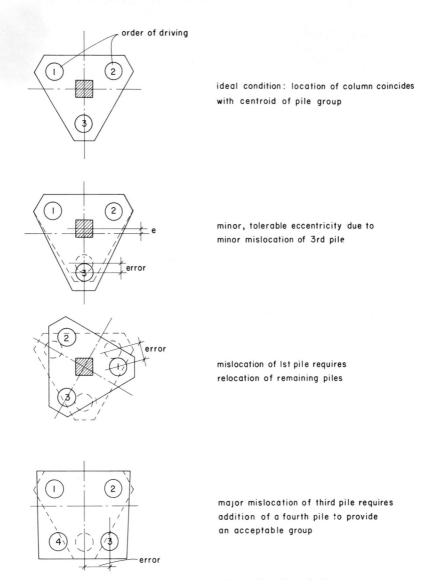

ideal condition: location of column coincides with centroid of pile group

minor, tolerable eccentricity due to minor mislocation of 3rd pile

mislocation of Ist pile requires relocation of remaining piles

major mislocation of third pile requires addition of a fourth pile to provide an acceptable group

FIGURE 4.3. Adjusting for mislocation of piles.

system and each individual type of element and process must be carefully analyzed when establishing the feasibility of alternatives.

Large diameter piers are usually excavated by direct digging, with the shaft walls lined and braced as the excavation proceeds. Once the pier excavation is completed, the lining is usually removed as the shaft is filled with concrete.

Placing of piers generally permits more precise control over the location of the shaft than that possible with piles. For this reason piers are usually not placed in clusters, or groups, but are merely increased or decreased in size as the load requirement varies. For drilled piers the shaft diameter may be held constant while the bells of individual piers are varied on the basis of the loads to be carried and/or the soil actually encountered upon excavation of the shaft to its final depth. Pier groups may be used when the structure to be carried is not a single column. Thus bearing walls, stair and elevator towers, and large items of equipment may be supported on several piers.

Although caps are not basically required for single piers, they are sometimes used. Caps may be used to eliminate the need for high accuracy of placement of anchor bolts or column reinforcing dowels at the time of the relatively rough foundation construction work. Caps are also used where a transition must be made between a very high strength concrete column and the usual low-strength pier shaft; thus the cap becomes similar in function to a pier for a footing.

4.3 Piles

Timber piles were used by ancient builders and are still in wide use today, primarily because of their low cost and widespread availability. However, industrialized countries also make use of steel and concrete for pile construction, especially where loads are high or rotting of timber piles is a potential problem. The major types of piles, in addition to timber, are steel, cast-in-place concrete and precast concrete.

As discussed in the previous section, the specific type of pile and the means for placing it are often quite specialized with respect to a particular contractor or locality. Thus the true relative feasibility or practicality of one system over another must be considered on a local basis. The following discussion deals with some of the general characteristics of the basic types of piles and driving methods and typical usage considerations for ordinary conditions.

Timber Piles. Timber piles consist of straight tree trunks, similar to those used for utility poles, that are driven with the small end down, primarily as friction piles. Their length is limited to that obtainable from the species of tree available. In most areas where timber is plentiful, lengths up to 50 or 60 ft are obtainable, whereas piles up to 80 or 90 ft may be obtained in some areas. The maximum driving force, and consequently the usable load, is limited by the problems of shattering either the leading point or the driven end. It is generally not possible to drive timber piles through very hard soil strata or through soil containing large rocks. Usable design working loads are typically limited to 50 to 60 k.

Decay of the wood is a major problem, especially where the tops of piles are above the ground-water line. Treatment with creosote will prolong the pile life but is only a delaying measure, not one of permanent protection. One technique is to drive the wood piles below the waterline and then build concrete piers on top of them up to the desired support level for the building.

For driving through difficult soils, or to end bearing, wood piles are sometimes fitted with steel points. This reduces the problem of damage at the leading point, but does not increase resistance to shattering at the driven end.

Because of their relative flexibility, long timber piles may be relatively easily diverted during driving, with the pile ending up in something other than a straight, vertical position. The smaller the pile group, the more this effect can produce an unstable structural condition. Where this is considered to be a strong possibility, piles are sometimes deliberately driven at an angle, with the outer piles in a group splayed out for increased lateral stability of the group. While not often utilized in buildings, this splaying out, called battering, of the outer piles is done routinely for foundations for isolated towers and bridge piers in order to develop resistance to lateral forces.

Timber piles are somewhat limited in their ability to accommodate to variations in driven length. In some situations the finished length of piles can only be approximated, as the actual driving resistance encountered establishes the required length for an individual pile. Thus the specific length of the pile to be driven may be either too long or too short. If too long, the timber pile can easily be cut off. However, if it is too short, it is not so easy to splice on additional length. Typically, the lengths chosen for the piles are quite conservatively long, with considerable cutting off tolerated in order to avoid the need for splicing.

Cast-in-Place Concrete Piles. Figure 4.4 shows various methods of installing concrete piles in which the shaft of the pile is cast in place in

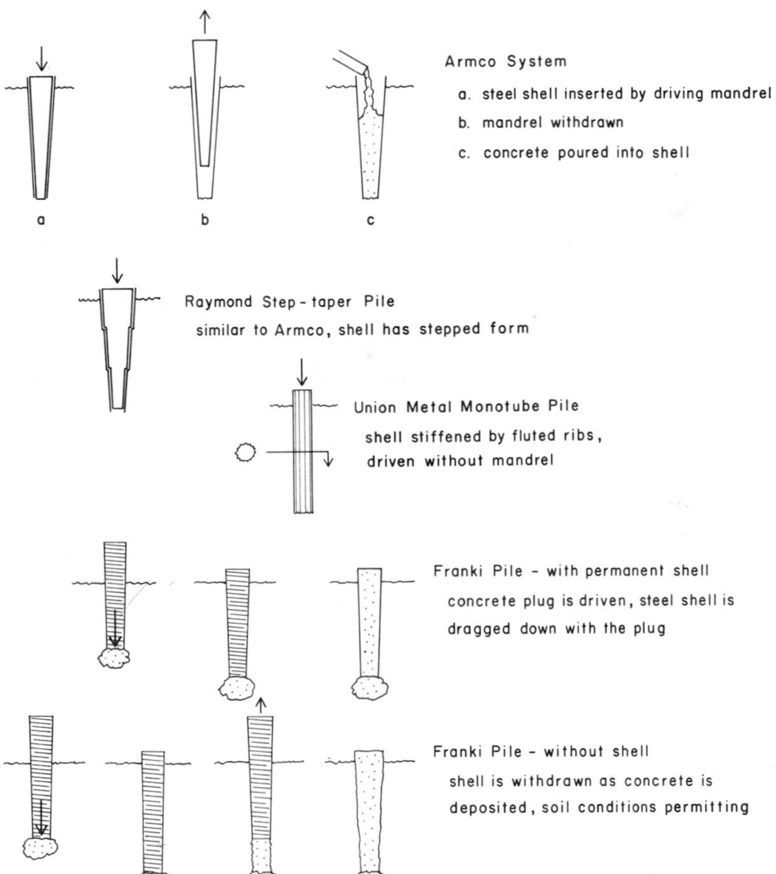
Armco System
a. steel shell inserted by driving mandrel
b. mandrel withdrawn
c. concrete poured into shell
a
b
c
Raymond Step-taper Pile
similar to Armco, shell has stepped form
Union Metal Monotube Pile
shell stiffened by fluted ribs, driven without mandrel
Franki Pile - with permanent shell
concrete plug is driven, steel shell is dragged down with the plug
Franki Pile - without shell
shell is withdrawn as concrete is deposited, soil conditions permitting

FIGURE 4.4. Examples of systems for producing cast-in-place concrete piles.

the ground. Most of these systems utilize materials or equipment produced by a particular manufacturer, who in some cases is also the installation contractor. As shown in the illustration, the systems are as follows:

Armco system. In this system a thin-walled steel cylinder is driven by inserting a heavy steel driving core, called a mandrel, inside the cylinder. The cylinder is then dragged into the ground as the mandrel is driven. Once in place, the mandrel is removed for reuse and the hollow cylinder is filled with concrete.

Raymond Step-Taper pile. This is similar to the Armco system in that a heavy core is used to insert a thin-walled cylinder into the ground. In this case the cylinder is made of spirally corrugated sheet steel and has a tapered vertical profile, both of which tend to increase the skin friction.

Union Metal Monotube pile. With this system the hollow cylinder is fluted longitudinally to increase its stiffness, permitting it to be driven without the mandrel. The fluting also increases the surface area, which tends to add to the friction resistance.

Franki pile with permanent steel shell. The Franki pile is created by depositing a mass of concrete into a shallow hole and then driving this concrete "plug" into the ground. Where a permanent liner is desired for the pile shaft, a spirally corrugated steel shell is engaged to the concrete plug and is dragged down with the driven plug. When the plug has arrived at the desired depth, the steel shell is then filled with concrete.

Franki pile without permanent shell. In this case the plug is driven without the permanent shell. If conditions require it, a smooth shell is used and is withdrawn as the concrete is deposited. The concrete fill is additionally rammed into the hole as it is deposited, which assures a tight fit for better friction between the concrete and the soil.

Both length and load range is limited for these systems, based on the size of elements, the strength of materials, and the driving techniques. The load range generally extends from timber piles at the lower end up to as much as 400 k for some systems.

Precast Concrete Piles. Some of the largest and highest load capacity piles have been built of precast concrete. In larger sizes these are usually hollow cylinders in order to reduce both the amount of material used and the weight for handling. These are more generally used for bridges and waterfront construction. A problem with these piles is establishing their precise in-place length. They are usually difficult to cut off as well as to splice. One solution is to produce them in modular lengths with a typical splice joint, which permits some degree of adjustment. The final finished top is then produced as a cast-in-place concrete cap.

In smaller sizes these piles are competitive in load capacity with those of cast-in-place concrete and steel. For deep water installations huge piles several hundred feet in length have been produced. These are floated into place and then dropped into position with their own dead weight ramming them home, since driving such a large element is not possible.

Steel Piles. Steel pipes and H-sections are widely used for piles, especially where great length or load capacity is required or where driving is difficult and requires excessive driving force. Although the piles themselves are quite expensive, their ability to achieve great length, their higher load capacity, and the relative ease of cutting or splicing them may be sufficient advantages to offset their price. As with timber piles of great length, their relative flexibility presents the problems of assuring exact straightness during driving.

As with timber piles, deterioration above the ground water level can be a problem. One solution is to cast a concrete jacket around the steel pile down to a point below the water level.

Table 4.1 summarizes information about some of the common types of piles for building construction. The data given are approximate and specific information about any type of pile should be obtained from local contractors or manufacturers.

When placed in groups, piles are ordinarily driven as close to each other as possible, primarily to reduce the structural requirements for the pile cap construction. The exact spacing allowable is related to the pile size and the driving technique. Ordinary spacings are 2 ft 6 in. for small timber piles and 3 ft for most other piles of the size range ordinarily used in building foundations.

TABLE 4.1. Information for Commonly Used Piles

Type of Pile	Usual Size Range: Diameter at Top (in.)	Usual Size Range: Length (ft)	Usual Load Capacity (k)
Wood	12 and up	50–80	60–100
Concrete, precast			
Solid	8–12	30–60	60–100
Hollow	No limit 18–36 common	Up to 200	80–200
Concrete, cast-in-place	15–36	40–100	60–150
Steel			
H-section	8–14 (nominal)	Up to 300	80–400
Pipe	10–36	Up to 200	100–400

Pile caps function much like column footings, and will generally be of a size close to that of a column footing for the same total load with a relatively high soil pressure. Pile layouts typically follow classical patterns, based on the number of piles in the group. Typical layouts are shown in Figure 4.5. Special layouts, of course, may be used for groups carrying bearing walls, shear walls, elevator towers, combined foundations for closely spaced columns, and other special situations.

Although the three-pile group is ordinarily preferred as the minimum for a column, the use of lateral bracing between groups may offer a degree of additional stability permitting the possibility of using a two-pile group, or even a single pile, for lightly loaded columns. This may extend the feasibility of using piles for a given situation, especially where column loads are less than that developed by even a single pile, which is not uncommon for single-story buildings of light construction and a low roof live load. Lateral bracing may be provided by foundation walls or grade beams or by the addition of ties between pile caps.

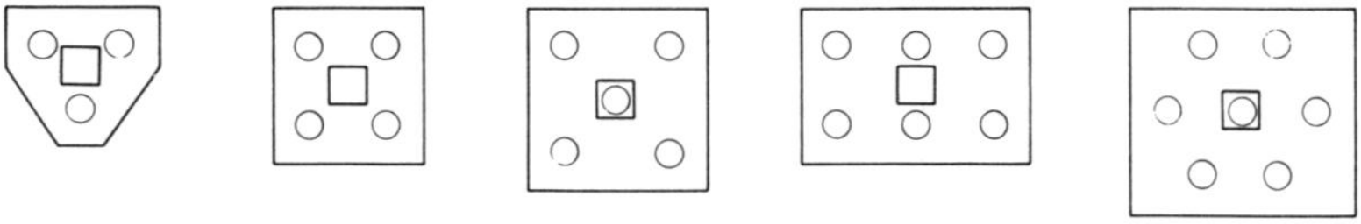

FIGURE 4.5. Typical pile arrangements and caps.

4.4 Piers

When loads are relatively light, the most common form of pier is the drilled-in pier consisting of a vertical round shaft and a bell-shaped bottom, as shown in Figure 4.6. When soil conditions permit, the pier shaft is excavated with a large auger-type drill similar to that used for large post holes and water wells. When the shaft has reached the desired bearing soil strata, the auger is withdrawn and an expansion element is inserted to form the bell. The decision to use such a foundation, the determination of the necessary sizes and details for the piers, and the development of any necessary inspection or testing during the construction should all be done by persons with experience in this type of construction.

This type of foundation is usually feasible only when a reasonably strong soil can be reached with a minimum-length pier. The pier shaft is usually designed as an unreinforced concrete column, although the

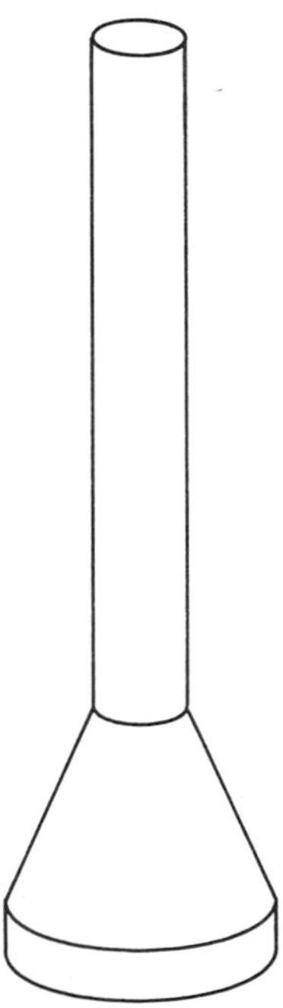

FIGURE 4.6. Typical form of the drilled pier with a belled bottom.

TABLE 4.2. Load Capacity of Drilled Piers

Shaft Diameter (ft)	Bell Diameter (ft)	Bell Bearing Capacity (k) and Minimum f'_c Required for Shaft (k/in²)[a] Allowable Soil Pressure for Bell (lb/ft²)			
		10,000	15,000	20,000	30,000
1.5	3	71	106	141	212
		2	2	2	3
	4.5	159	239	318	477
		2.5	3.5	4.5	6.5
2	4	126	188	251	377
		2	2	2	3
	6	283	424	565	848
		2.5	3.5	4.5	6.5
3	6	283	424	565	848
		2	2	2	3
	9	636	954	1272	1909
		2.5	3.5	4.5	6.5
4	8	503	754	1005	1508
		2	2	2	3
	12	1131	1696	2262	3393
		2.5	3.5	4.5	6.5
5	10	785	1178	1571	2356
		2	2	2	3
	15	1767	2651	3534	5301
		2.5	3.5	4.5	6.5
6	12	1131	1696	2262	3393
		2	2	2	3
	18	2545	3817	5089	7634
		2.5	3.5	4.5	6.5
7	14	1539	2309	3079	4618
		2	2	2	3
	21	3464	5195	6927	10391
		2.5	3.5	4.5	6.5

[a] Based on allowable $f_c = 0.30\, f'_c$.

upper part of the shaft is often provided with some reinforcement. This is done to give the upper part of the pier some additional resistance to bending caused by lateral forces or column loads that are slightly eccentric from the pier centroid.

The usual limit for the bell diameter is three times the shaft diameter. With this as an upper limit, actual bell diameters are sometimes determined at the time of drilling on the basis of field tests performed on the soil actually encountered at the bottom of the shaft.

Table 4.2 gives the capacities for various-size belled piers for a range of allowable soil pressures. Loads are given for bell diameters of two and three times the shaft diameter. In addition to the load capacity based on the area of the bell, the table gives the corresponding minimum value for the concrete in the shaft. This minimum value for f_c' is based on a matching of the pier strength to the limit for the bell bearing capacity and the use of an allowable stress in the shaft of $f_c = 0.3 f_c'$. While this is a stress limit that is commonly used, individual building codes may establish lower limits.

One of the advantages of drilled piers is that they may usually be installed with a higher degree of control on the final position of the pier tops than is possible with driven piles. It thus becomes more feasible to consider the use of a single pier for the support of a column load. For the support of walls, shear walls, elevator pits, or groups of closely spaced columns, however, it may be necessary to use clusters or rows of piers. The minimum spacing for such groups of piers is essentially limited by the bell diameters, with some spacing between edges of bells required to assure that the drilling of one pier does not disturb previously installed piers.

4.5 Lateral, Uplift, and Moment Effects on Deep Foundations

The resistance to horizontal forces, vertically directed upward forces, and moments presents special problems for deep foundation elements. Whereas a bearing footing has no potential for the development of tension between the structure and the ground, both piles and piers have considerable capacity for uplift resistance. On the other hand, the sliding friction that constitutes a major resistance to horizontal force by a bearing footing is absent with deep foundation elements. The following discussion deals with some of the problems of designing deep foundations for force effects other than the primary one of vertically directed downward load.

Lateral Force Resistance. Resistance to horizontal force at the top of both piles and piers is very poor in most cases. The relatively narrow profile offers little contact surface for the development of passive soil pressure. In addition, the process of installation generally causes considerable disturbance of the soil around the top of the foundation elements. Although there are procedures for determination of the resistance that can be developed by the passive pressure, this resistance is seldom utilized as the major force-resolving effect in building design. In addition to the low magnitude of the force that can be developed, there is the problem of considerable movement due to the soil deformation, which constitutes a dimensional distortion that few buildings can tolerate.

The usual method for resolving horizontal forces with deep foundation systems is to transfer the forces from the piles or piers to other parts of the building construction. In most cases this means the use of ties and struts to transfer the forces to grade walls or basement walls that offer considerable surface area for the development of passive soil pressures. This procedure is also used with bearing footings when the footings themselves are not capable of the total force resistance required.

For large freestanding structures without grade walls, the horizontal force resistance of pile groups is usually developed through the use of some piles driven at an angle. These so-called *battered piles* are capable of considerable horizontal force resistance, in both compression and tension. This practice is common in the construction of foundations for large towers, bridge abutments, and so on, but is rarely utilized in building construction, where the load sharing described previously is usually the more feasible design option.

Uplift Resistance. Friction piles and large piers have considerable resistance to upward forces. If skin friction is truly the main resistive force that constitutes the pile capacity to sustain downward load, then it should also resist force in the opposite direction. An exception is the pile with a tapered form, which will have slightly higher resistance in one direction. Another exception is the unreinforced concrete pile, which has considerably more resistance to compressive stresses than it has to tensile stresses on the pile shaft.

The combined weight of the shaft and bell of a drilled pier offers a considerable potential force for resistance to upward loads. In addition, if the bell is large in diameter, considerable soil pressure can be developed against the upward withdrawal of the pier from the ground. Finally, if the shaft is long and the hole is without a permanent steel casing, some skin friction will be developed. These effects can be combined into a major resistance to upward force. However, the concrete shaft must be heavily reinforced, or prestressed, in order to develop the full potential resistance of the pier.

End-bearing piles usually have much less resistance to uplift in comparison to their compressive load capacities. However, in spite of the existence of relatively weak upper soil strata, there is usually some potential for skin friction resistance to withdrawal of the pile.

Uplift resistance of piles must usually be established by field load tests if they are to be relied on for major design loads. For large piers, and also for very large individual piles, this may not be feasible, because of the load magnitudes involved. However, in many design situations the uplift forces required are less than the limiting capacity of the deep elements, in which case it is sometimes acceptable to rely on a very conservative calculation of the uplift capacity.

The potential for uplift resistance on the part of deep foundation elements is an element of design that is not present with bearing foundations. Thus design for moments or for actual tension anchorage may be approached differently with deep foundations. In the case of the shear wall examples illustrated in Section 3.13, for example, the concerns for overturn and maximum soil pressure must be resolved without any reliance on tensile resistance between the footing and the soil. Thus the footing length and the length of the large grade beam must both be increased until the necessary relationships are developed through the use of weight and compression stress on the soil. With deep foundations this structure could possibly be reduced by shortening the grade beam and relying on the tensile capacity of the deep foundation elements.

Moment Resistance. Piles and piers are seldom deliberately designed to develop moments. Although any element strong enough to function as a pile or pier inevitably possesses some bending strength, it is generally considered desirable to design for an ideal condition of axial load only. However, the unavoidable inaccuracies inherent in the construction processes sometimes make it impractical to assure the perfect alignment inferred by the assumption of axial loading. Thus the true evaluation of any structural element that is primarily intended to carry axial load usually includes some consideration of the probability of accidental moments produced by eccentricity of the load.

The larger the diameter of a pile or pier, the larger the potential eccentricity that can be tolerated with a significant load magnitude. If the construction process can reasonably assure a dimensional control of placement error within this tolerable eccentricity, then potential accidental misalignment may not be a problem. This is the primary judgment to be made in justifying the use of a single pile or pier for a single concentrated load, such as that from an individual column.

Figure 4.7 illustrates the general effects of construction inaccuracy that are possible when two compression members interact to transfer load. At the left is shown the two types of eccentricity, produced by misalignment or lack of straightness in the members. Two problems must be dealt with in this situation. First, the moments produced in the members must be added to the compression force for a combined stress analysis. As shown in the center of Figure 4.7 the eccentricity produces

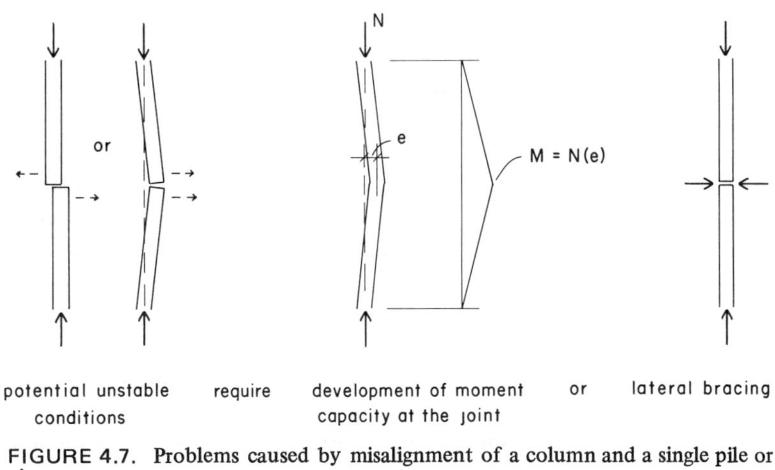

FIGURE 4.7. Problems caused by misalignment of a column and a single pile or pier.

a maximum moment equal to the product of the compression force times the eccentricity distance. If this force/moment interaction and the resulting combined stresses can be tolerated by the members and the joint between them, the situation may not be critical from this point of view. However, the second consideration is the lateral movement of the elements. If this is unresisted, the situation becomes one of inherent instability, regardless of the load magnitude or sizes of the members. It is necessary, therefore, to provide some means for resistance of the lateral movement. One possibility is to use a joint between the members that has a moment capability, simulating the situation of a single continuous member. If this joint is capable of developing the moment of N times e, then the lateral movement will be resisted by the bending stiffness of the members. However, the better technique is that shown at the right in Figure 4.7, where the lateral movement is prevented by separate constraining forces. If these forces are sufficient, the movement and instability problem is eliminated, although the force/moment interaction and combined stress problem remains.

For piles and piers there are generally two basic means for developing the lateral restraining forces shown in Figure 4.7. Passive horizontal soil pressure will offer some restraint to the pile or pier. In most cases, however, this will require some soil deformation to develop a significant force. This movement will generally further aggravate the load eccentricity problem and is thus not very desirable. The other possible constraining force, and the one to be preferred, is that offered by elements of the construction that connect to the tops of the piles or piers. These may be foundation walls, grade beams, floor slabs, or ties and struts. The relative stiffness of these elements will virtually prevent any movement of the joint at the top of the pile or pier.

A second moment problem occurs when groups of piles must accept a force/moment combination, similar to that considered for moment-resistive footings in Section 3.11. As shown in Figure 4.8, the same concepts that were developed in the discussion of the effects of load eccentricity on a footing apply when considering similar effects on pile groups. The pile group has a kern limit, similar to that for the footing plan section. Load within this limit will produce a distribution of loads to the piles as shown in Case 1 in Figure 4.8, with all piles in compression, although some are more highly stressed. Load eccentricity outside the kern will produce tension in the piles on the opposite edge of the

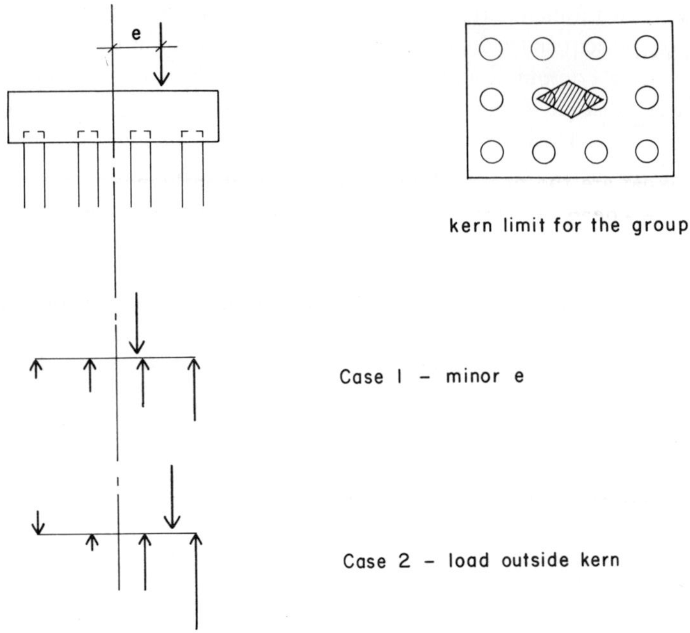

FIGURE 4.8. Effects of an eccentric load on a pile group.

group, a situation that may be possible in this case, although it was not with the footing.

Study Aids

Words and Terms. Using the glossary and the text of the preceding chapter for reference, review the meaning of the following words and terms.

Belled pier.
Caisson.
Drilled pier.
End-bearing pile.
Friction pile.
Hardpan.

Pier.

Pile cap.

Questions

1. What are the principal considerations that influence a decision to use a deep foundation instead of a shallow bearing foundation?
2. Why are piles usually placed in groups?
3. What soil conditions make the installation of piles or drilled piers difficult?
4. How are horizontal forces on piles and piers usually resisted?
5. What makes it possible to use single, isolated piers, whereas piles are usually placed in groups?
6. What is the principal cause of the deterioration of the tops of steel and timber piles?

5

Design for Horizontal Forces

Horizontal force effects are considered in the design of many types of foundation elements. Discussion of these effects is included in the general development of design considerations for a number of elements presented in various other sections of this book. The material in this chapter deals with the general problems of forces that are horizontally developed in soil masses and with the design of three types of structures for which horizontal force effects are a major concern: basement walls, freestanding retaining walls, and abutments.

5.1 Horizontal Force Effects

There are a number of situations involving horizontal forces in soils. The three major ones of concern in foundation design are the following:

Active soil pressure. Active soil pressure originates with the soil mass; that is, it is pressure exerted *by* the soil on something, such as the outside surface of a basement wall.

Passive soil pressure. Passive soil pressure is exerted *on* the soil, for example that developed on the side of a footing when horizontal forces push on the footing.

Friction. Friction is the sliding effect developed between the soil and the surface of some object in contact with the soil. To develop friction there must be some pressure between the soil and the contact fact of the object.

The development of all of these effects involves a number of different stress mechanisms and structural behaviors in the various types of soils. A complete treatment of these topics is beyond the scope of this book, and the reader is referred to other references for such a discussion. (See *Foundation Engineering*, Ref. 1, or *Introductory Soil Mechanics and Foundations*, Ref. 2.) The discussion that follows will explain the basic phenomena and illustrate the use of some of the simple techniques for design utilizing data and procedures from existing codes.

Active Soil Pressure. The nature of active soil pressure can be visualized by considering the situation of an unrestrained vertical cut in a soil mass, as shown in Figure 5.1. In most soils such a cut, as shown at (*a*), will not stand for long. Under the action of various influences, primarily gravity, the soil mass will tend to move to a form as shown at (*b*), producing an angled profile, rather than the vertical face.

There are two general forces involved in the change from the vertical to the sloped cut profile. The soil near the top of the cut tends to simply drop vertically under its own weight. The soil near the bottom of the cut tends to bulge out horizontally from the cut face, being squeezed by the soil mass above it. Another way to visualize this movement is to consider that the whole moving soil mass tends to rotate with respect to a slip plane such as that indicated by the heavy dashed line at (*c*) in Figure 5.1.

If a restraining structure of some type is introduced at the vertical

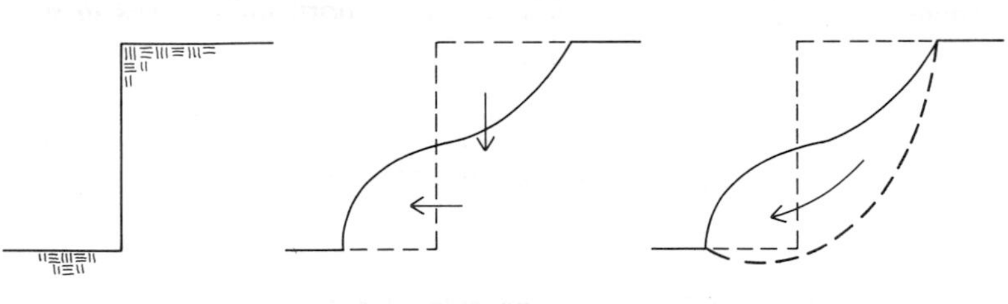

FIGURE 5.1. Typical failure of a vertical cut.

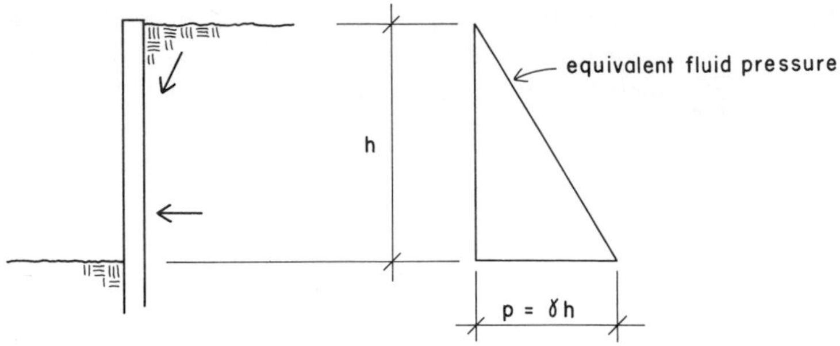

FIGURE 5.2. Lateral pressure on a vertical restraining structure.

face, the forces exerted on it by the soil will tend to be those involved in the actions illustrated in Figure 5.1. As shown in Figure 5.2, the soil mass near the top of the cut will have combined vertical and horizontal effects. The horizontal component of this action will usually be minor, since the mass at this location will move primarily downward, and may develop significant friction on the face of the restraining structure. The soil mass near the bottom of the cut will exert primarily a horizontal force, very similar to that developed by a liquid in a tank. In fact, the most common approach to the design for this situation is to assume that the soil acts as a fluid with a unit density of some percentage of the soil weight, and to consider a horizontal pressure that varies with the height of the cut, as shown in Figure 5.2.

The simplified equivalent fluid pressure assumption is in general most valid when the retained soil is a well-drained sandy soil. If the soil becomes saturated, the water itself will increase the horizontal pressure, and the buoyancy effect will tend to reduce the resistance to the slip-plane rotation of the soil mass as illustrated in Figure 5.1. If the soil contains a high percentage of silt or clay, the simple linear pressure variation as a function of the height is quite unrealistic.

In addition to considerations of the soil type and the water content, it is sometimes necessary to deal with a surcharge on the retained soil mass. As shown in Figure 5.3, the two common situations involving a surcharge are when the soil mass is loaded by some added vertical force, such as the wheel load of a vehicle, and when the ground profile is not flat, producing the effect of raising its level with respect to the top of the restraining structure. The surcharge tends to increase the pressure

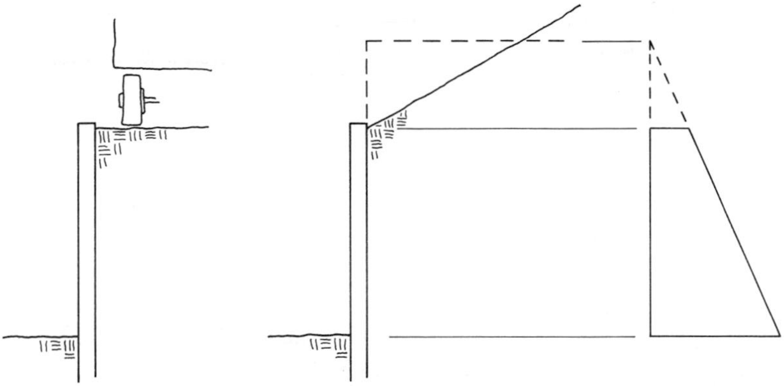

FIGURE 5.3. Surcharge effects on a restraining structure.

near the top of the wall. When the equivalent fluid pressure method is used, the usual procedure is to consider the top of the soil mass (and the zero stress point for the fluid pressure) to be above the top of the structure. This results in a general overall increase in the fluid pressure, which is somewhat conservative in the case of the wheel load, whose effect tends to diminish with distance below the contact point. When handled as fluid pressure, the surcharge is sometimes simply visualized either as an increase in the assumed density of the equivalent fluid, or as the addition of a certain height of soil mass above the top of the retaining structure.

Passive Soil Pressure. Passive soil pressure is visualized by considering the effect of pushing some object through the soil mass. If this is done in relation to a vertical cut, as shown in Figure 5.4, the soil mass will tend to move inward and upward, causing a bulging of the ground sur-

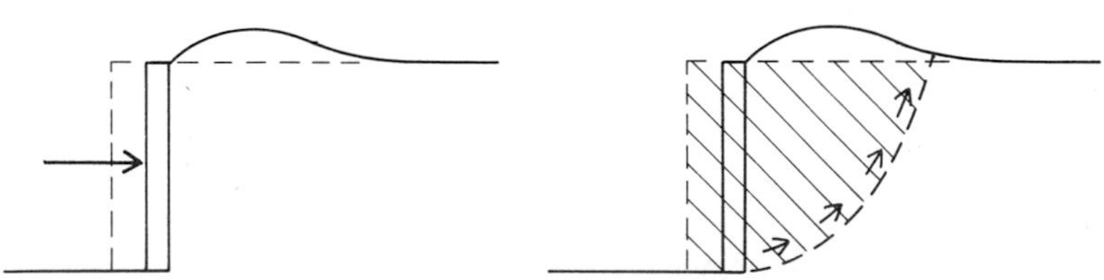

FIGURE 5.4. Development of passive soil resistance.

face behind the cut. If the slip-plane type of movement is assumed, the action is similar to that of active soil pressure, with the directions of the soil forces simply reversed. Since the gravity load of the upper soil mass is a useful force in this case, passive soil resistance will generally exceed active pressure for the same conditions.

If the analogy is made to the equivalent fluid pressure, the magnitude of the passive pressure is assumed to vary with depth below the ground surface. Thus for structures whose tops are at ground level, the pressure variation is the usual simple triangular form as shown in the left-hand illustration in Figure 5.5. If the structure is buried below the ground surface, as is the typical case with footings, the surcharge effect is assumed and the passive pressures are correspondingly increased.

As with active soil pressure, the type of soil and the water content will have some bearing on development of stresses. This is usually accounted for by giving values for specific soils to be used in the equivalent fluid pressure analysis, as illustrated in the examples that follow.

Soil Friction. The potential force in resisting the slipping between some object and the soil depends on a number of factors, including the following principal ones:

Form of the contact surface. If a smooth object is placed on the soil there will be a considerable tendency for it to slip. Our usual concern

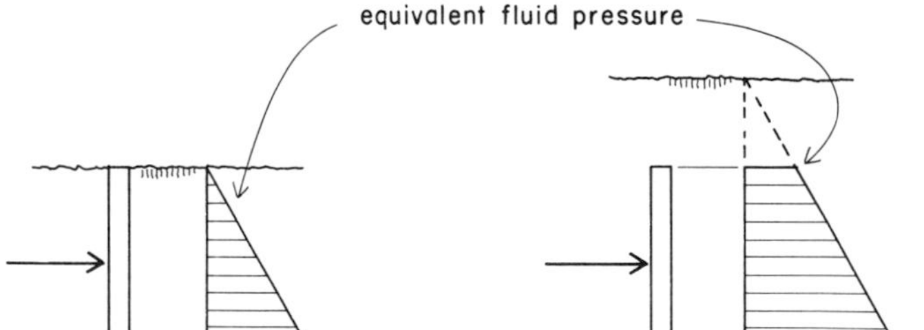

FIGURE 5.5. Passive soil pressure development—with and without surcharge.

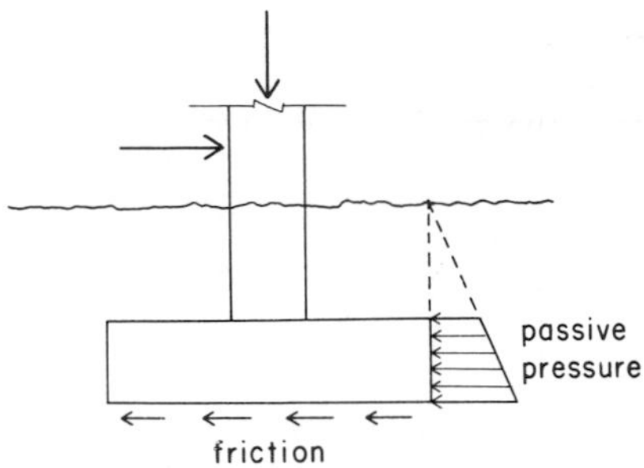

FIGURE 5.6. Development of resistance to horizontal force by a footing.

is for a contact surface created by pouring concrete directly onto the soil, which tends to create a very nonsmooth, intimately bonded surface.

Type of soil. The grain size, grain shape, relative density, and water content of the soil are all factors that will affect the development of soil friction. Well-graded, dense, angular sands and gravels will develop considerable friction. Loose, rounded, saturated, fine sand and soft clays will have relatively low friction resistance. For sand and gravel the friction stress will be reasonably proportional to the compressive pressure on the surface, up to a considerable force. For clays, the friction tends to be independent of the normal pressure, except for the minimum pressure required to develop any friction force.

Pressure distribution on the contact surface. When the normal surface pressure is not constant, the friction will also tend to be nonuniform over the surface. Thus, instead of an actual stress calculation, the friction is usually evaluated as a total force in relation to the total load generating the normal stress.

Friction seldom exists alone as a horizontal resistive force. Foundations are ordinarily buried with their bottoms some distance below the ground surface. Thus pushing the foundation horizontally will also usually result in the development of some passive soil pressure, as shown in Figure 5.6. Since these are two totally different stress mechanisms, they will actually not develop simultaneously. Nevertheless, the usual practice is to assume both forces to be developed in opposition to the total horizontal force on the structure.

5.2 Basement Walls

Basement walls are usually used as earth-retaining structures. They also typically function as bearing walls for the vertical gravity loads of the building, and sometimes serve the spanning or load distributing functions as described in Section 3.1. Thus the stress conditions in basement walls may be quite complex, requiring analysis for several combinations of loading for accurate determination of critical situations.

For their earth-retaining function, basement walls ordinarily span vertically between levels of support. For a single-story basement the support at the bottom of the wall is provided by the edge of the concrete basement floor slab, and the support at the top of the wall is provided by the first-floor structure of the building. If an active soil pressure of the fluid type is assumed, the load and structural actions for the wall will be as shown in Figure 5.7(*a*) when the ground level is at the top of

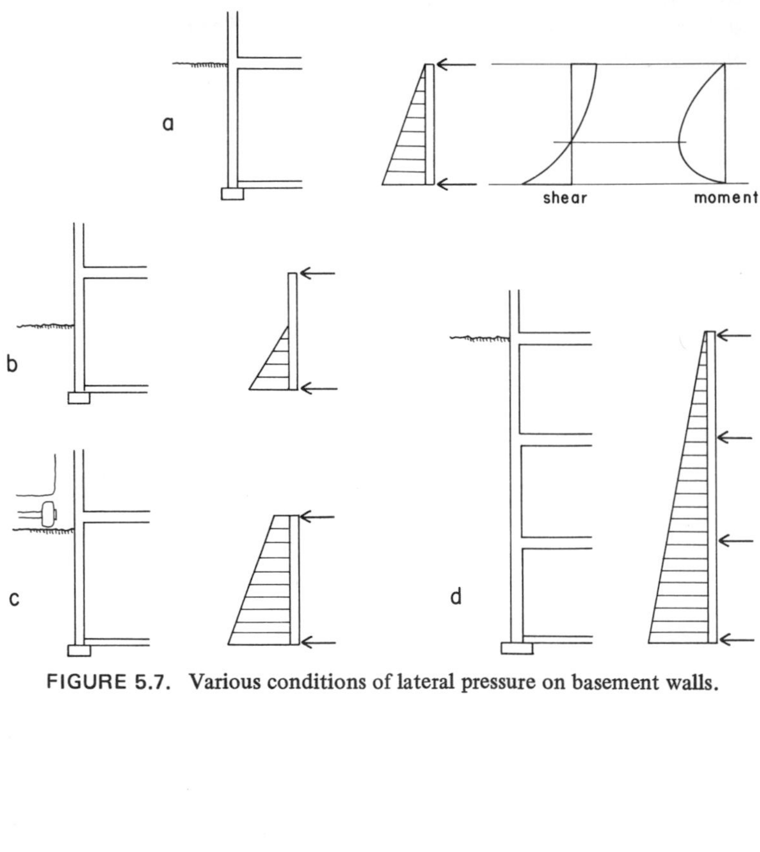

FIGURE 5.7. Various conditions of lateral pressure on basement walls.

the wall. When the ground level is below the top of the wall, the pressure is as shown in (*b*). A surcharge load at the edge of the building will increase the pressure as shown in (*c*). If the basement is multilevel, the wall will usually function as a multiple span element for the pressure as shown in (*d*).

For buildings of light construction with relatively short basement walls, the walls are often built of masonry or concrete without reinforcing. Precise analysis of many such walls would show a condition of overstress by current codes, but long experience with few failures is generally accepted as a valid reason for continuing the practice. Nevertheless, when the wall spans vertically more than 10 ft or is subjected to surcharge effects, we recommend the use of walls of reinforced concrete or reinforced fully grouted masonry. And, of course, when the wall has additional structural tasks to perform, they should be included in the design stress analysis.

The following example illustrates the design of a simple vertically spanning basement wall of reinforced concrete. The design conforms to general requirements of the ACI Code for cover of reinforcing and minimum reinforcing in both vertical and horizontal directions. The example illustrates the basis for determination of the entries in Table 5.1.

Example 16 Design of a Reinforced Concrete Basement Wall

Design data and criteria:

Concrete design strength: $f'_c = 3000$ psi.
Reinforcing: $f_s = 20{,}000$ psi.
Active soil pressure: 30 lb/ft^2 per ft of depth below surface.
Surcharge: 300 lb/ft^2 on ground surface (equivalent to 3 ft of additional soil).

The wall is as shown in Figure 5.8. The wall spans vertically between the lateral supports provided by the basement floor and the structure supported on top of the wall. We assume the span to be the clear height of the wall. For ease of pouring the concrete we recommend limiting the wall height to approximately 15 times the thickness. For the 10 in. thick wall this limits the height to $12\frac{1}{2}$ feet. Subtracting for the base-

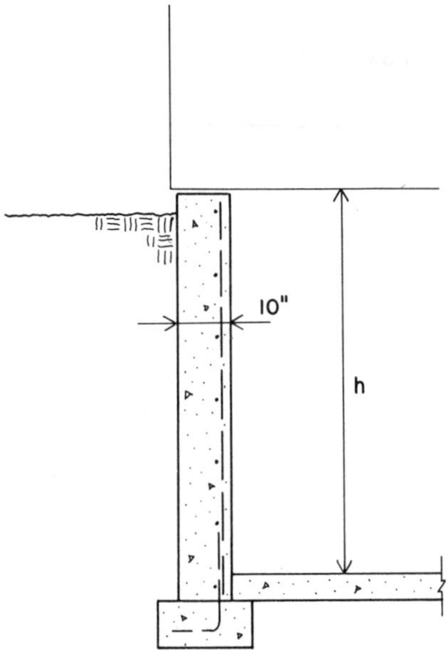

FIGURE 5.8. The basement wall—Example 16.

ment floor, we will therefore consider the maximum clear height to be 12 ft, and will design the wall for this span.

The ACI Code (Ref. 10) recommends the following minimum reinforcing for the wall:

Horizontal:

$$A_s = 0.0025\, A_g = (0.0025)(120) = 0.30 \text{ in.}^2/\text{ft}$$

Vertical:

$$A_s = 0.0015\, A_g = (0.0015)(120) = 0.18 \text{ in.}^2/\text{ft}$$

Thus we would use the following as minimum reinforcing for the wall, unless structural calculations indicate larger areas.

Horizontal: No. 5 bars at 12 in., $A_s = 0.31$ in.^{2}/ft.
Vertical: No. 4 bars at 13 in., $A_s = 0.185$ in.^{2}/ft.

The beam action of the wall is shown in Figure 5.9. Because of the surcharge, the pressure at the top of the wall is (3)(30) = 90 lb/ft^2 using

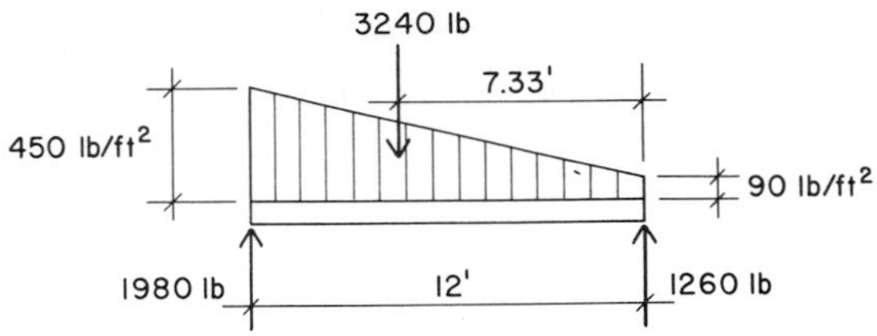

FIGURE 5.9. Loading condition for the spanning wall–Example 16.

the equivalent fluid pressure method. This pressure increases at the rate of 30 lb/ft² per additional ft of depth, to the maximum value of 450 lb/ft² at the bottom of the wall. In terms of the span and the unit of the pressure variation, the maximum moment produced for this loading will be approximately

$$M = 0.064\,ph^3 + 0.375\,ph^2$$

and for this example:

$$M = (0.064)(30)(12)^3 + (0.375)(30)(12)^2 = 3318 + 1620 = 4938 \text{ lb-ft}$$

This moment may be compared to the balanced moment capacity in order to consider the concrete bending stress and the relative values to be used for k and j.

$$\text{Balanced } M = Rbd^2 = (226)(12)(9)^2(\tfrac{1}{12}) = 18{,}306 \text{ lb-ft}$$

which indicates that concrete stress is not critical and the section will be considerably underreinforced, permitting a conservative guess of 0.90 or higher for j.

We have used the value of 9 in. for the effective depth of the reinforced concrete section, which assumes the reinforcing to be placed with the minimum clearance of $\frac{3}{4}$ in. on the inside face of the wall. With these approximations for j and d, the area of steel required is determined as follows:

$$A_s = \frac{M}{f_s jd} = \frac{(4938)(12)}{(20{,}000)(0.9)(9)} = 0.366 \text{ in.}^2/\text{ft}$$

which could be furnished with

No. 5 bars at 10 in., $A_s = 0.37$ in.²/ft.
No. 6 bars at 14 in., $A_s = 0.38$ in.²/ft.

TABLE 5.1. Reinforced Concrete Basement Walls

Wall Height h (ft)	Wall Thickness t (in.) 8	10	12	
	Vertical Reinforcing (Bar Size–Spacing in Inches) $\frac{3}{4}$ in. clear of inside of wall			2 in. Clear of Outside of Wall
8	4–14	4–13	4–18	4–18
9	4–11	4–13	4–16	4–18
10	5–12	5–16	5–18	4–18
11		5–13	5–15	4–18
12		5–10	5–12	4–18
13	not recommended		5–10	4–18
14	$h > 15\,t$		6–11	4–18
Horizontal reinforcing	5–15	5–12	5–10 (one face) 4–13 (both faces)	

Note from Figure 5.9 that the required support force at the bottom of the wall is 1980 lb. Unless there is considerable dead load on the top of the wall, it will probably be necessary to require that the basement floor slab be placed before the backfill is deposited against the wall.

It is possible, of course, that with a considerable vertical load the wall may have a critical combined stress or load/moment interaction. It is also possible that the vertical load may be sufficiently off-center of the wall to produce significant moment, which may add to the bending due to the soil pressure. These conditions, plus any others due to grade beam action, load distributing, and so forth, should be considered in the full design of a basement wall.

Table 5.1 gives reinforcing recommendations for some concrete walls, as determined by the procedures illustrated in the preceding example. Figure 5.10 shows the conditions assumed for the walls.

There are many potential detail requirements for basement walls. Although we do not intend to make recommendations for all of the construction requirements for these walls, it is well to be aware of considerations such as the following:

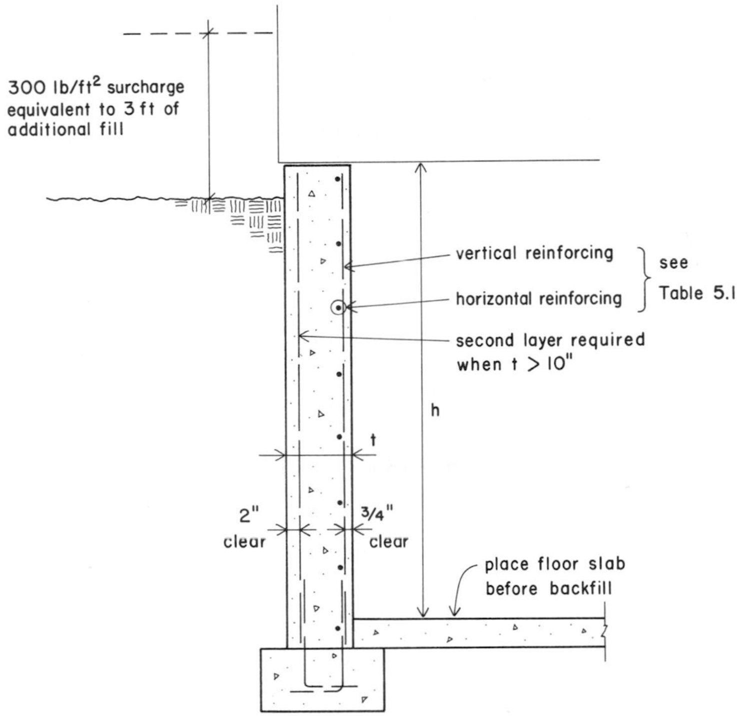

FIGURE 5.10. Conditions and recommendations for the reinforced concrete basement walls in Table 5.1.

Need for waterproofing. The design of concern for water penetration depends on the ground water situation. When the ground water level is well below the bottom of the wall, the need is for what is technically called *dampproofing*. This is usually accomplished in two stages. First the wall itself is made as impervious as possible through the use of good mortar joints, vibrating or otherwise working concrete to eliminate air bubbles, segregation, and so forth, and careful detailing and construction to avoid developing cracks in the wall. Then, if necessary, an asphaltic compound is applied to the outside of the wall up to the ground level. If the wall must actually be *waterproof*, that is, it must resist actual hydrostatic pressure of standing water, it must be treated in a manner essentially similar to that for a flat roof, with a waterproofing membrane of some kind applied

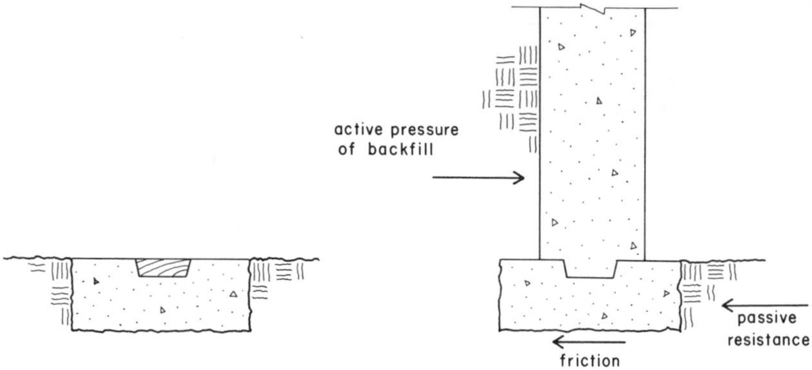

FIGURE 5.11. Use of keyways in wall footings.

to the wall surface. In addition, in the latter case joints in the wall must be watertight. This usually involves the use of some combination of inserted waterstops and applied joint sealing compound. For masonry walls, the exterior surface is sometimes finished with a coating of cement plaster, although this practice is now less common with the advent of quality dampproofing compounds.

Need for temporary bracing. In some situations there may be compelling reasons for not pouring the basement floor slab until some later stage of the building construction. If this results in a need for placing the backfill against the walls prior to placing the basement floor slab, the bottoms of the walls must be braced adequately for the lateral earth pressure. The placing of a keyway slot in the top of the wall footing, as shown in Figure 5.11, is an old common practice for providing some such support for the wall. Actually, if the wall is reinforced with vertical bars, the action of the dowels plus the friction due to the wall weight may be sufficient to develop considerable force resistance, and the key slot is really superfluous. And, of course, if the floor slab is placed before backfilling, the key slot is without any function.

5.3 Retaining Walls

Strictly speaking, any wall that sustains significant lateral soil pressure is a retaining wall. However, the term is usually used with reference to a so-called *cantilever retaining wall*, which is a freestanding wall without

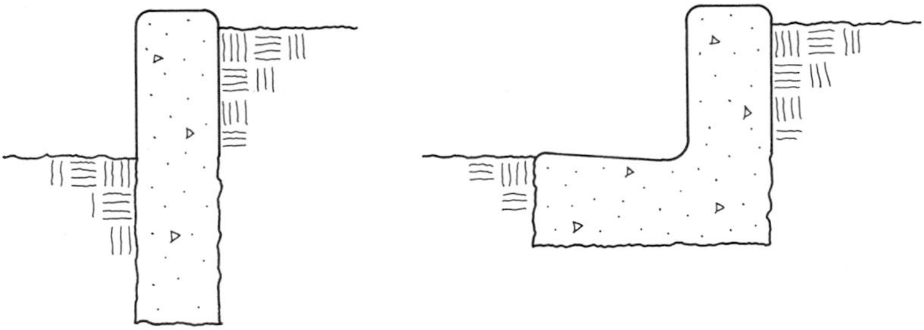

FIGURE 5.12. Typical forms used for low curbs.

lateral support at its top. For such a wall the major design consideration is for the actual dimension of the ground level difference that the wall serves to facilitate. The range of this dimension establishes some different categories for the retaining structure, as follows:

Curbs. Curbs are the shortest freestanding retaining structures. The two most common forms are as shown in Figure 5.12, the selection being made on the basis of whether or not it is necessary to have a gutter on the low side of the curb. Use of these structures is typically limited to grade level changes of about 2 ft or less.

Short retaining walls. Vertical walls up to about 10 ft in height are usually built as shown in Figure 5.13. These consist of a concrete or

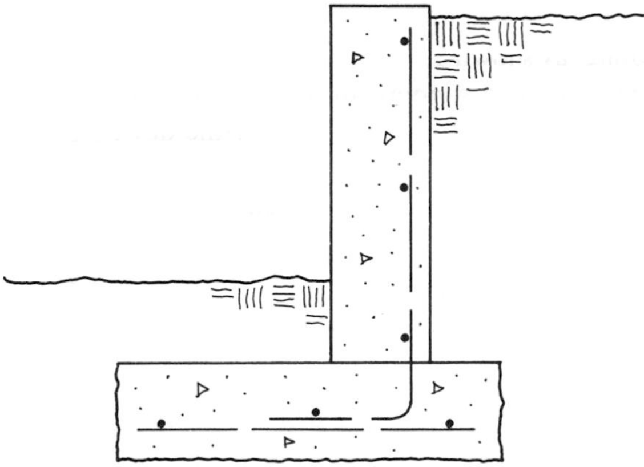

FIGURE 5.13. Typical form of a short cantilever retaining wall.

masonry wall of uniform thickness. The wall thickness, footing width and thickness, vertical wall reinforcing, and transverse footing reinforcing are all designed for the lateral shear and cantilever bending moment plus the vertical weights of the wall, footing, and earth fill.

Tall retaining walls. As the wall height increases it becomes less feasible to use the simple construction shown in Figure 5.13. The overturning moment increases sharply with the increase in height of the wall. For very tall walls one modification used is to taper the wall thickness. This permits the development of a reasonable cross section for the high bending stress at the base without an excessive amount of concrete. However, as the wall becomes really tall, it is often necessary to consider the use of various bracing techniques, as shown in the other illustrations in Figure 5.14.

The design of tall retaining walls is beyond the scope of this book. They should be designed with a more rigorous analysis of the active soil pressure than that represented by the simplified equivalent fluid

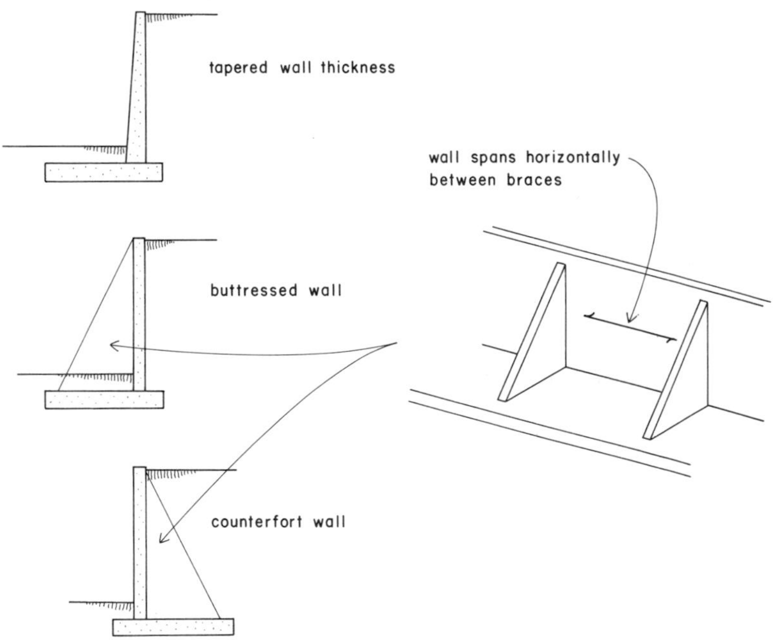

FIGURE 5.14. Forms used for tall cantilever retaining walls.

stress method. In addition, the magnitudes of forces in the reinforced concrete elements of such walls indicate the use of strength design methods, rather than the less-accurate working stress methods.

Under ordinary circumstances it is reasonable to design relatively short retaining walls by the equivalent fluid pressure method and to use the working stress method for the design of the elements of the wall. The following example illustrates this simplified method of design.

EXAMPLE 17 SHORT RETAINING WALL

The wall is to be of reinforced concrete with the profile shown in Figure 5.15. Design data and criteria are as follows:

Active soil pressure: 30 lb/ft^2 per ft of height.
Soil weight: assumed to be 100 lb/ft^3.
Maximum allowable soil pressure: 1500 lb/ft^2.
Concrete strength: $f_c' = 3000$ psi.
Allowable tension on reinforcing: 20,000 psi.

The loading condition used to analyze the stress conditions in the wall is shown in Figure 5.16.

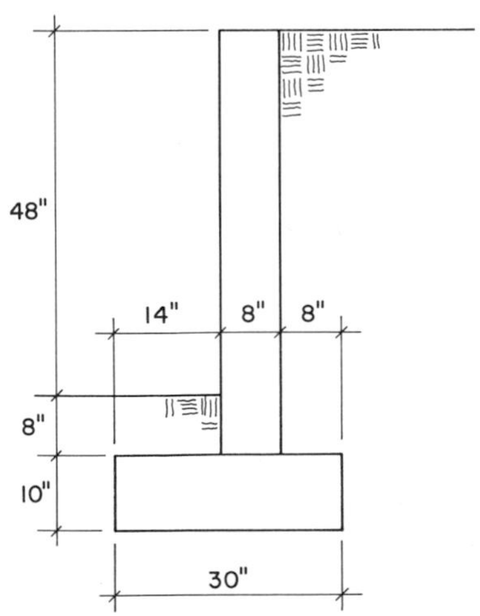

FIGURE 5.15. Form of the retaining wall–Example 17.

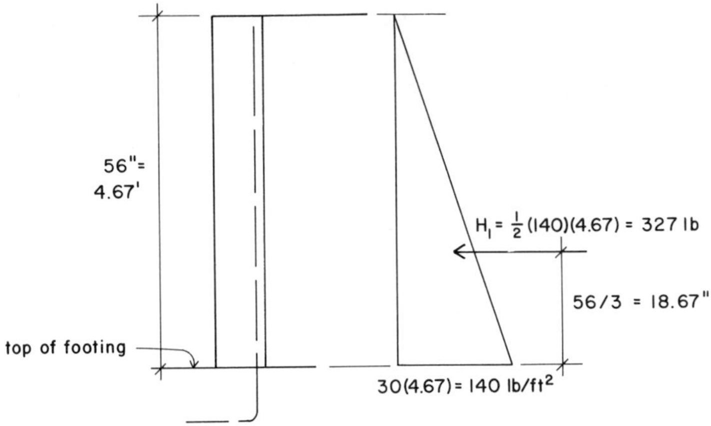

FIGURE 5.16. Soil pressure loading for the wall analysis.

Maximum lateral pressure:

$$p = (30)(4.667 \text{ ft}) = 140 \text{ lb/ft}^2$$

Total horizontal force:

$$H_1 = \frac{(140)(4.667)}{2} = 327 \text{ lb}$$

Moment at base of wall:

$$M = (327)\left(\tfrac{56}{3}\right) = 6104 \text{ lb-in.}$$

For the wall we assume an approximate effective d of 5.5 in. The tension reinforcing required for the wall is thus

$$A_s = \frac{M}{f_s jd} = \frac{6104}{(20{,}000)(0.9)(5.5)} = 0.061 \text{ in.}^2/\text{ft}$$

This may be provided by using No. 3 bars at 20 in. centers, which gives an actual A_s of 0.066 in.^{2}/ft. Since the embedment length of these bars in the footing is quite short, they should be selected conservatively and should have hooks at their ends for additional anchorage.

The loading condition used to investigate the soil stresses and the stress conditions in the footing is shown in Figure 5.17. In addition to the limit of the maximum allowable soil bearing pressure, it is usually

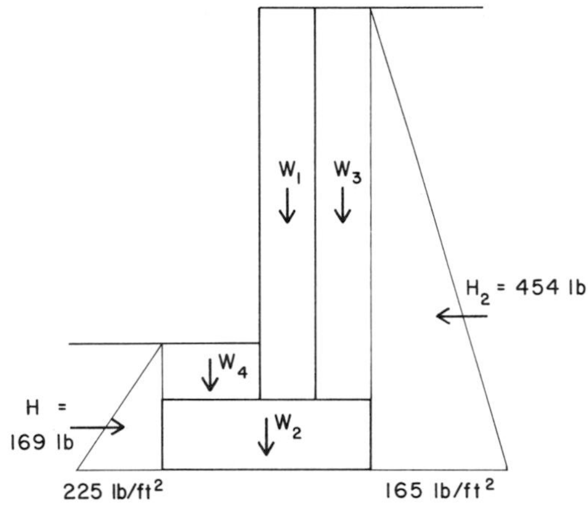

FIGURE 5.17. Loading for the footing analysis–Example 17.

required that the resultant vertical force be kept within the kern limit of the footing. The location of the resultant force is therefore usually determined by a moment summation about the centroid of the footing plan area, and the location is found as an eccentricity from this centroid.

Table 5.2 contains the data and calculations for determination of the location of the resultant force that acts at the bottom of the footing. The position of this resultant is found by dividing the net moment by the sum of the vertical forces, as follows:

$$e = \frac{5793}{1167} = 4.96 \text{ in.}$$

TABLE 5.2. Determination of the Eccentricity of the Resultant Force

Force (lb)		Moment Arm (in.)	Moment (lb-in.)
H_2	454	22	+9988
w_1	466	3	-1398
w_2	312	0	0
w_3	311	11	-3421
w_4	78	8	+624
	Σ_w = 1167 lb		Net moment = +5793 lb-in

For the rectangular footing plan area the kern limit will be $\frac{1}{6}$ of the footing width, or 5 in. The resultant is thus within the kern, and the combined soil stress may be determined by the stress formula as follows:

$$p = \frac{N}{A} \pm \frac{M}{S}$$

in which: N is the total vertical force.
A is the plan area of the footing.
M is the net moment about the footing centroid.
S is the section modulus of the rectangular footing plan area, which is determined as follows:

$$S = \frac{bh^2}{6} = \frac{(1)(2.5)^2}{6} = 1.042 \text{ ft}^3$$

The limiting maximum and minimum soil pressures are thus determined as follows:

$$p = \frac{N}{A} \pm \frac{M}{S} = \frac{1167}{2.5} \pm \frac{5793/12}{1.042} = 467 \pm 463$$

$$= 930 \text{ lb/ft}^2 \text{ maximum and } 4 \text{ lb/ft}^2 \text{ minimum}$$

Since the maximum stress is less than the established limit of 1500 lb/ft^2, vertical soil pressure is not critical for the wall. For the horizontal force analysis the procedure varies with different building codes. The criteria given in this example for soil friction and passive resistance are those in the *Uniform Building Code* (Ref. 9; see also the appendix) for ordinary sandy soils. This code permits the addition of these two resistances without modification. Using this data and technique, the analysis is as follows:

Total active force: 454 lb, as shown in Figure 5.17.

Friction resistance [(friction factor) (total vertical dead load)]:

$$(0.25)(1167) = 292 \text{ lb}$$

Passive resistance: 169 lb, as shown in Figure 5.17.

Total potential resistance:

$$292 + 169 = 461 \text{ lb}$$

Since the total potential resistance is greater than the active force, the wall is not critical in horizontal sliding.

As with most wall footings, it is usually desirable to select the footing thickness to minimize the need for tension reinforcing due to bending. Thus shear and bending stresses are seldom critical, and the only footing stress concern is for the tension reinforcing. The critical section for bending is at the face of the wall, and the loading condition is as shown in Figure 5.18. The trapezoidal stress distribution produces the resultant force of 833 lb, which acts at the centroid of the trapezoid, as shown in the illustration. Assuming an approximate depth of 6.5 in. for the section, the analysis is as follows:

Moment:

$$M = (833)(7.706) = 6419 \text{ lb-in}$$

Required area:

$$A_s = \frac{M}{f_s jd} = \frac{6149}{(20{,}000)(0.9)(6.5)} = 0.055 \text{ in.}^2/\text{ft}$$

This requirement may be satisfied by using No. 3 bars at 24 in. centers. For ease of construction it is usually desirable to have the same spacing for the vertical bars in the wall and the transverse bars in the footing. Thus in this example the No. 3 bars at 20 in. centers previously selected

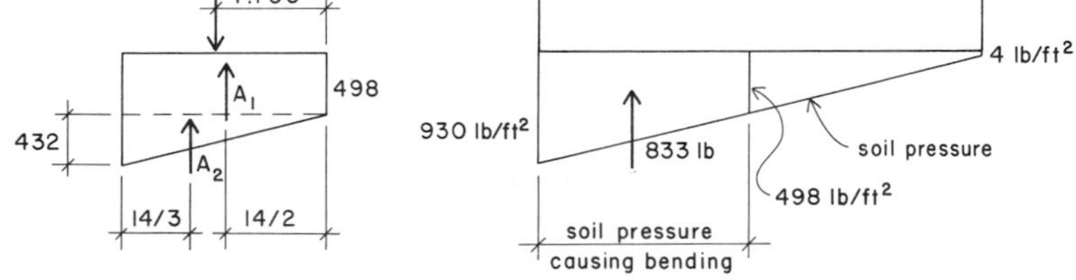

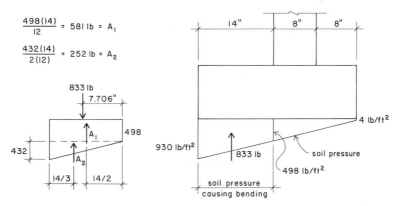

FIGURE 5.18. Analysis for the vertical soil pressure effects–Example 17.

for the wall would probably also be used for the footing bars. The vertical bars can then be held in position by wiring the hooked ends to the transverse footing bars.

Although bond stress is also a potential concern for the footing bars, it is not likely to be critical as long as the bar size is relatively small (less than a No. 6 bar or so).

Reinforcing in the long direction of the footing should be determined in the same manner as for ordinary wall footings. As discussed in Section 3.2, we recommend a minimum of 0.15 percent of the cross section. For the 10 in. thick and 30 in. wide footing this requires

$$A_s = (0.0015)(300) = 0.45 \text{ in.}^2$$

We would therefore use three No. 4 bars with a total area of (3) (0.2) = 0.6 in.^{2}.

In most cases designers consider the stability of a short cantilever wall to be adequate if the potential horizontal resistance exceeds the active soil pressure and the resultant of the vertical forces is within the kern of the footing. However, the stability of the wall is also potentially questionable with regard to the usual overturn effect. If this investigation is considered to be necessary, the procedure is as follows.

The loading condition is the same as that used for the soil stress analysis and shown in Figure 5.17. As with the vertical soil stress analysis, the force due to passive soil resistance is not used in the moment calculation, since it is only a potential force. For the overturn investigation the moments are taken with respect to the toe of the footing. The calculation of the overturning and dead load restoring moments are shown in Table 5.3. The safety factor against overturn is determined as

$$SF = \frac{\text{restoring moment}}{\text{overturning moment}} = \frac{20{,}686}{9988} = 2.07$$

The overturning effect is usually not considered to be critical as long as the safety factor is at least 1.5.

Table 5.4 gives design data for short reinforced concrete retaining walls varying in height from 2 to 6 ft. Table data has been developed using the procedures illustrated in Example 17. Details and criteria for the walls are shown in Figure 5.19. Note that the illustration shows two necessary conditions. The first concerns the profile of the ground sur-

TABLE 5.3. Analysis for Overturning Effect

Force (lb)		Moment Arm (in.)	Moment (lb-in.)
Overturn:			
H_2	454	22	9988
Restoring moment:			
w_1	466	18	8388
w_2	312	15	4680
w_3	311	26	8086
w_4	78	7	546
		Total:	20,686 lb-in.

face behind the wall. If this has a significant slope there will be an increase in the active soil pressure similar to that due to a surcharge. Table designs are based on consideration of an essentially flat profile, although a very minor slope (up to 5 : 1, as shown) will not cause significant increase in pressure. The second requirement is that care be taken

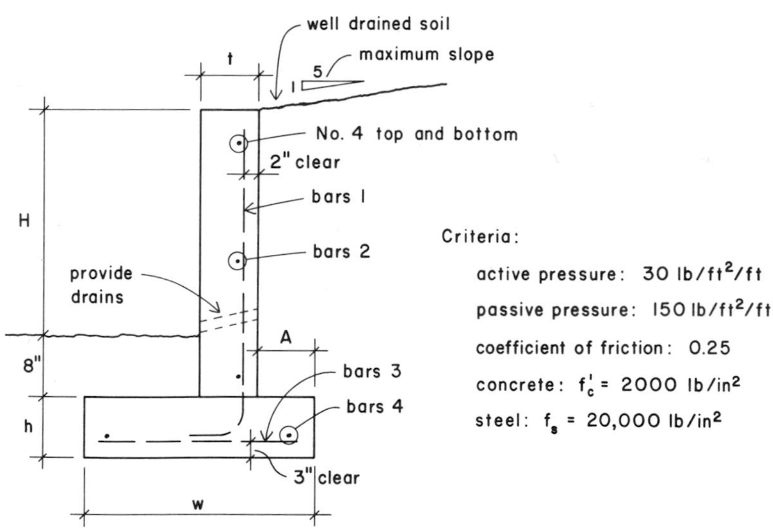

FIGURE 5.19. Conditions and recommendations for the concrete retaining walls in Table 5.4.

TABLE 5.4. Short Concrete Retaining Walls[a]

Wall Height H (ft)	Wall and Footing Dimensions (ft-in.)				Reinforcing				Actual Maximum Soil Pressure (lb/ft^2)
	w	h	t	A	1	2	3	4	
2	1–6	0–6	0–6	0–4	No. 3 at 30	–	–	2 No. 3	750
3	2–0	0–8	0–6	0–6	No. 3 at 24	1 No. 4	–	2 No. 4	800
4	2–6	0–10	0–8	0–8	No. 3 at 20	2 No. 4	No. 3 at 20	3 No. 4	950
5	3–4	1–0	0–9	1–1	No. 4 at 24	3 No. 4	No. 4 at 24	4 No. 4	900
6	4–4	1–3	0–10	1–4	No. 4 at 18	4 No. 4	No. 4 at 18	4 No. 5	925

[a] See Figure 5.19 for reference.

TABLE 5.5. Short Masonry Retaining Walls[a]

Wall Height H (ft)	Wall		Footing			Reinforcing			Actual Maximum Soil Pressure (lb/ft^2)
	Nominal t (in.)	Assumed Weight lb/ft^2 of Wall Surface	w (in.)	h (in.)	A (in.)	1	2	3	
2	6	55	18	6	4	No. 3 at 48	–	2 No. 3	550
2.67	6	55	22	6	6	No. 3 at 32	–	2 No. 3	600
3.33	8	75	27	8	8	No. 4 at 48	No. 4 at 48	2 No. 4	700
4	8	75	32	10	10	No. 4 at 32	No. 4 at 32	3 No. 4	850
4.67	8	75	40	12	12	No. 4 at 24	No. 3 at 24	4 No. 4	850
5.33	10	95	48	14	15	No. 4 at 24	No. 4 at 24	5 No. 5	825
6	10	95	56	16	18	No. 5 at 24	No. 4 at 24	5 No. 5	850

[a] See Figure 5.20 for reference.

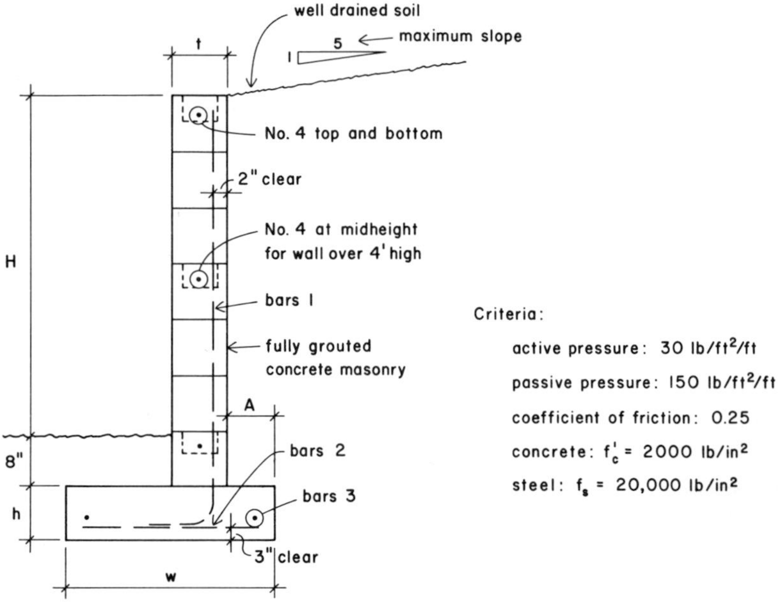

FIGURE 5.20. Conditions and recommendations for the masonry retaining walls in Table 5.5.

to avoid the possibility of highly saturated soil behind the wall. This should be avoided by using a reasonably permeable fill and by placing drains in the wall as shown.

Table 5.5 gives design data for short reinforced masonry retaining walls varying in height from 2 to 6 ft. Table data have been developed using procedures essentially similar to those in Example 17. Details and criteria for the walls are shown in Figure 5.20. Wall thicknesses are based on typical nominal block sizes. For the determination of the wall weight it is assumed that the blocks are of lightweight aggregate and have all voids filled with concrete.

5.4 Abutments

The support of some types of structures, such as arches, gables, and shells, often requires the resolution of both horizontal and vertical forces. When this resolution is accomplished entirely by the supporting

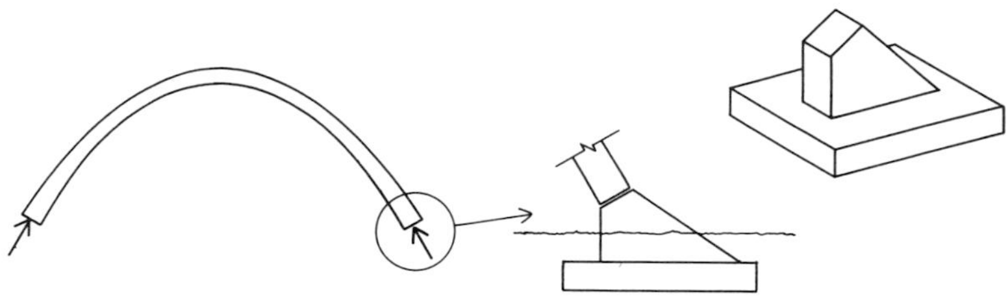

FIGURE 5.21. Abutment for an arch.

foundation element, the element is described as an abutment. Figure 5.21 shows a simple abutment for an arch, consisting of a rectangular footing and an inclined pier. The design of such a foundation has three primary concerns, as follows:

Resolution of the vertical force. This consists of assuring that the vertical soil pressure does not exceed the maximum allowable value for the soil.

Resolution of the horizontal force. If the abutment is freestanding, resolution of the horizontal force means the development of sufficient soil friction and passive horizontal pressure.

Resolution of the moment effect. In this case the aim is usually to keep the resultant force as close as possible to the centroid of the footing plan area. If this is truly accomplished—that is, $e = 0$—there will literally be no moment effect on the footing itself.

Figure 5.22 shows the various forces that act on an abutment such as that shown for the arch in Figure 5.21. The active forces consist of the load and the weights of the pier, the footing, and the soil above the footing. The reactive forces consist of the vertical soil pressure, the horizontal friction on the bottom of the footing, and the passive horizontal soil pressure against the sides of the footing and pier. The dashed line in the illustration indicates the path of the resultant of the active forces; the condition shown is the ideal one, with the path coinciding with the centroid of the footing plan area at the bottom of the footing.

If the passive horizontal pressure is ignored, the condition shown in Figure 5.22 will result in no moment effect on the bottom of the footing and a uniform distribution of the vertical soil pressure. If the passive

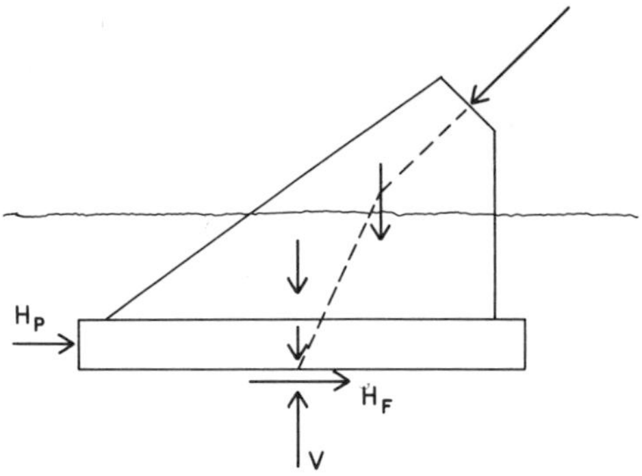

FIGURE 5.22. Loading condition for an abutment.

horizontal pressure is included in the force summation, the resultant path will move slightly to the right of the footing centroid. However, for the abutment as shown, the resultant of the passive pressure will be quite close to the bottom of the footing, so that the error is relatively small.

If the pier is tall and the load is large with respect to the pier weight or is inclined at a considerable angle from the vertical, it may be necessary to locate the footing centroid at a considerable distance horizontally from the load point at the top of the pier. This could result in a footing of greatly extended length if a rectangular plan form is used. One device that is sometimes used to avoid this is to use a T-shape, or other form, that results in a relocation of the centroid without excessive extension of the footing. Figure 5.23 shows the use of a T-shape footing for such a condition.

When the structure being supported is symmetrical, such as an arch with its supports at the same elevation, it may be possible to resolve the horizontal force component at the support without relying on soil stresses. The basic technique for accomplishing this is to tie the two opposite supports together, as shown in Figure 5.24, so that the horizontal force is resolved internally (within the structure) instead of externally (by the ground). If this tie is attached at the point of contact between the structure and the pier, as shown in Figure 5.24, the net load delivered to the pier is simply a vertical force, and the pier and footing could theoretically be developed in the same manner as that for

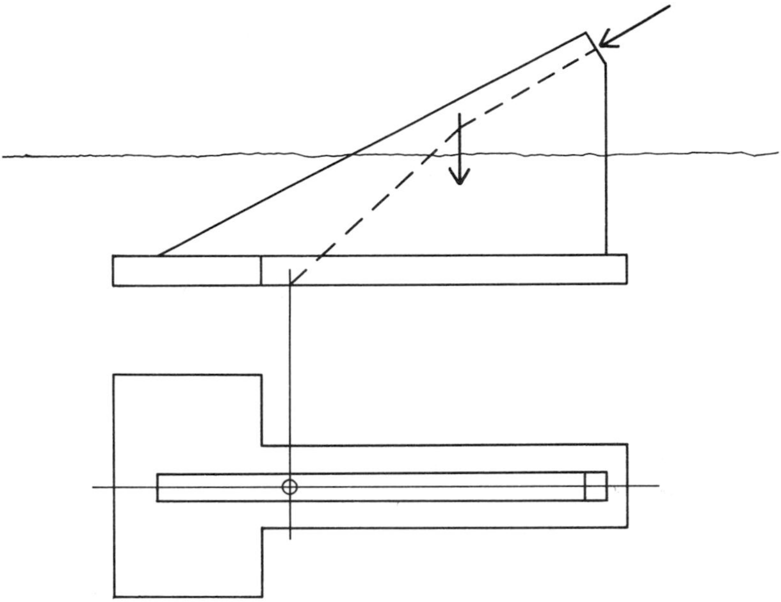

FIGURE 5.23. Use of a T-shaped footing for a large abutment.

a truss or beam without the horizontal force effect. However, since either wind or seismic loading will produce some horizontal force on the supports, the inclined pier is still the normal form for the supporting structure. The position of the footing, however, would usually be established by locating its centroid directly below the support point, as shown in the figure.

For practical reasons it is often necessary to locate the tie, if one is used, below the support point for the structure. If this support point is

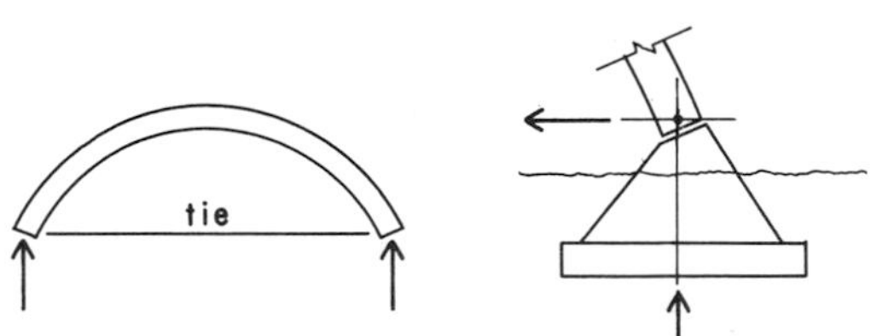

FIGURE 5.24. Abutment for a tied arch.

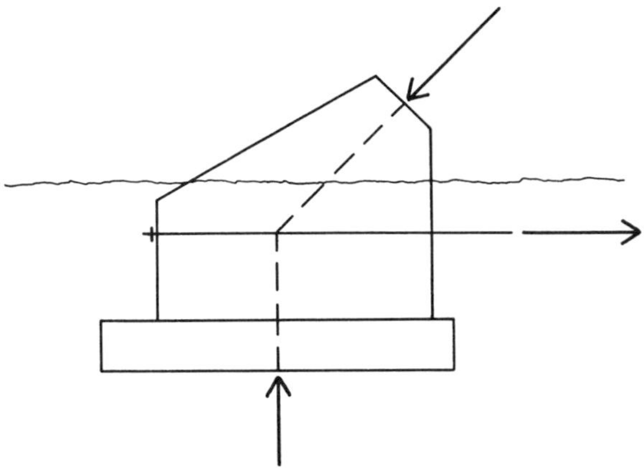

FIGURE 5.25. Abutment with a subgrade tie.

above ground, as it usually is, the existence of the tie aboveground is quite likely to interfere with the use of the structure. A possible solution in this problem is to move the tie down to the pier, as shown in Figure 5.25. In this case the pier weight is added to the load to find the proper location for the footing centroid.

When the footing centroid must be moved a considerable distance from the load point, it is sometimes necessary to add another element to the abutment system. Figure 5.26 shows a structure in which a large grade beam has been inserted between the pier and the footing. The main purpose of this element is to develop the large shear and bending resistance required by the long cantilever distance between the ends of the pier and footing. In the example, however, it also serves to provide for the anchorage of the tie. Because of this location of the tie, the weights of both the pier and grade beam would be added to the load to find the proper location of the footing centroid. In this way the heavy grade beam further assists the footing by helping to move the centroid closer to the pier and reducing the cantilever distance.

When the horizontal component of the load is large and it is not possible to use a tie, additional passive resistance may be developed by adding a crosswall perpendicular to the pier, as shown in Figure 5.27. A consideration that must be made in this case is that the resultant of the passive pressure will move up to some significant distance above

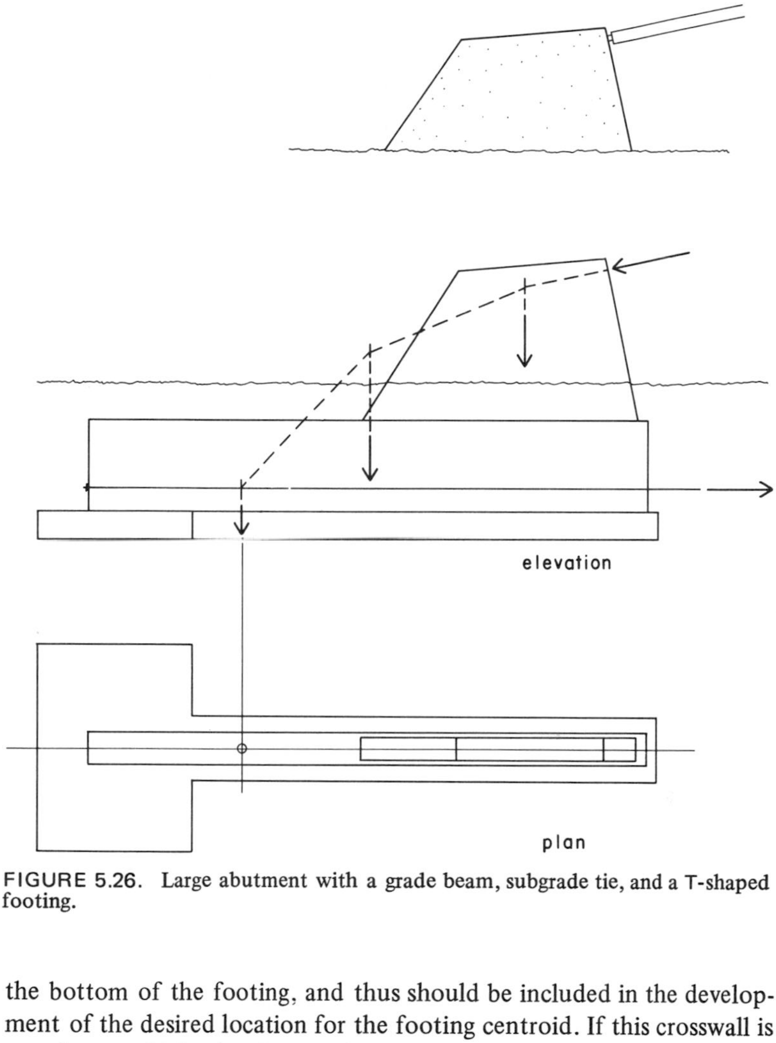

FIGURE 5.26. Large abutment with a grade beam, subgrade tie, and a T-shaped footing.

the bottom of the footing, and thus should be included in the development of the desired location for the footing centroid. If this crosswall is very long or high, the shear and moment produced by the cantilevering action of the crosswall beyond the sides of the pier may be quite significant. A possible solution in this case is to provide a horizontal slab between the pier and crosswall, as shown in dashed line profile in the isometric view in Figure 5.27. If there is a useful function for this slab,

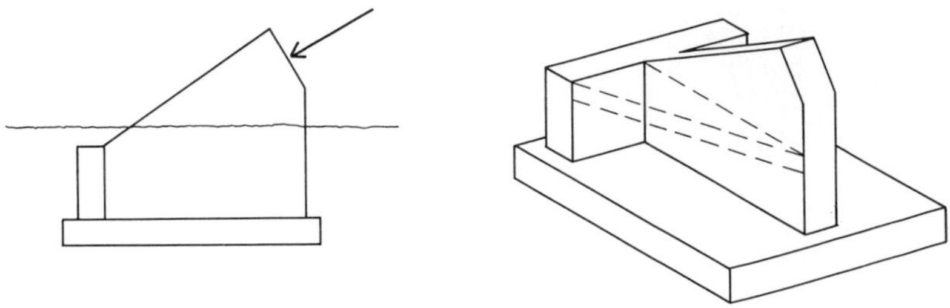

FIGURE 5.27. Use of a crosswall for increased development of passive soil pressure.

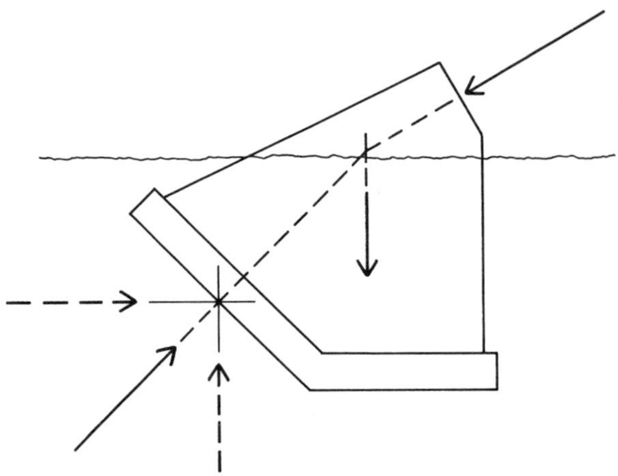

FIGURE 5.28. Use of an inclined footing for an abutment.

it may be placed aboveground; otherwise it may simply be buried along with the rest of the foundation.

A variation on the footing-plus-crosswall combination is shown in Figure 5.28. Here an inclined footing is used to develop both vertical and horizontal soil pressure. The crosswall is eliminated and the footing merely changes from horizontal to inclined along its length.

Study Aids

Words and Terms. Using the glossary and the text of the preceeding chapter for reference, review the meaning of the following words and terms.

Abutment.
Active lateral pressure.
Backfill.
Cantilever retaining wall.
Curb.
Dampproofing.
Equivalent fluid pressure.
Freestanding wall.
Friction.
Kern.
Key.
Overturning effect.
Passive lateral pressure.
Retaining wall.
Surcharge.
Waterproofing.

Questions

1. What is the difference between active lateral pressure and passive lateral pressure?
2. How is frictional resistance determined on:
   a. sand and gravel?
   b. clay?
3. How is lateral soil pressure determined by the equivalent fluid pressure method?
4. Why is it common to require that the basement floor slab be placed before backfill is deposited against the outside of a basement wall?

Problems

1. Select an entry in the table for reinforced concrete basement walls (Table 5.1) and verify the data from the table by performing the calculations illustrated in Example 16.

2. Select an entry in the table for short retaining walls (Table 5.4) and, using the illustrations in Example 17, do the following:

   a. Determine the actual maximum soil pressure.

   b. Determine the safety factor against overturn.

   c. Determine whether the footing has adequate resistance to sliding.

   d. Verify the data in the table.

6

Special Foundation Problems

The following sections deal with some of the special, although ordinary, problems that occur in building foundation design, in addition to the general problems of wall and column footings and retaining and foundation walls.

6.1 Paving Slabs

Sidewalks, driveways, and basement floors are typically produced by depositing a relatively thin coating of concrete directly on the ground surface. While the basic construction process is simple, a number of factors must be considered in developing details and specifications for a paving slab.

Thickness of the Slab. Pavings vary in thickness from a few inches (for residential basement floors) to several feet (for airport landing strips). Although more strength is implied by a thicker slab, thickness alone does not guarantee a strong pavement. Of equal concern is the reinforcement provided and the character of the sub-base on which the concrete is poured. The minimum slab thickness commonly used in building floor slabs is $3\frac{1}{2}$ inches. This relates specifically to the actual dimension of a nominal wood 2-by-4, and simplifies forming of the edges of a slab pour. Following the same logic, the next size jump would be to a $5\frac{1}{4}$ inch thickness, which is the dimension of a nominal 2-by-6.

The $3\frac{1}{2}$ in. thick slab is usually considered adequate for interior floors not subjected to wheel loadings or other heavy structural demand. At this thickness, usually provided with very minimal reinforcing, the slab has relatively low resistance to bending and shear effects of concentrated loads. Thus walls, columns, and heavy items of equipment should be provided with separate footings.

The $5\frac{1}{4}$ in. thick slab is adequate for heavier live loads and for light vehicular wheel loads. It is also strong enough to provide support for light partitions, so that some of the extra footing construction can be eliminated.

For heavy truck loadings, for storage warehouses, and for other situations involving very heavy loads—especially concentrated ones—thicker pavements should be used, although thickness alone is not sufficient, as mentioned previously.

Reinforcing. Thin slabs are ordinarily reinforced with welded wire mesh. The most commonly used meshes are those with a square pattern of wires—typically 4 or 6 in. spacings—with the same wire size in both directions. This reinforcing is generally considered to provide only for shrinkage and temperature effects and to add little to the flexural strength of the slab. The minimum mesh, commonly used with the $3\frac{1}{2}$ in. slab, is a 6X6 10/10, which denotes a mesh with No. 10 wires at 6 in. on center in each direction. For thicker slabs the wire gage should be increased or two layers of mesh should be used.

Small-diameter reinforcing bars are also used for slab reinforcing, especially with thicker slabs. These are generally spaced at greater distances than the mesh wires and must be supported during the pouring operation. Unless the slab is actually designed to span, as is discussed in Section 6.2, this reinforcing is still considered to function primarily for shrinkage and temperature stress resistance. However, since cracking in the exposed top surface of the slab is usually the most objectionable, specifications usually require the reinforcing to be kept some minimum distance from the top of the slab.

Sub-base. The ideal sub-base for floor slabs is a well-graded soil, ranging from fine gravel to coarse sand with a minimum of fine materials. This material can usually be compacted to a reasonable density to provide a good structural support, while retaining good drainage properties to avoid moisture concentrations beneath the slab. Where ground water conditions are not critical, this base is usually simply wetted down before pouring the concrete and the concrete is deposited directly on the sub-base. The wetting serves somewhat to consolidate the sub-base and to reduce the bleeding out of the water and cement from the bottom of the concrete mass.

To further reduce the bleeding-out effect, or where moisture penetration is more critical, a lining membrane is often used between the slab and the sub-grade or base. This usually consists of a 6 mil plastic sheet or a laminated paper-plastic-fiber glass product that possesses considerably more tear resistance and that may be desirable where the construction activity is expected to increase this likelihood.

Joints. Building floor slabs are usually poured in relatively small units, in terms of the horizontal dimension of the slab. The main reason is to control shrinkage cracking. Thus a full break in the slab, formed as a joint between successive pours, provides for the incremental accumulation of the shrinkage effects. Where larger pours are possible or more desirable, control joints are used. These consist of tooled or sawed joints that penetrate some distance down from the finished top surface.

Surface Treatment. Where the slab surface is to serve as the actual wearing surface, the concrete is usually formed to a highly smooth surface by troweling. This surface may then be treated in a number of ways, such as brooming it to make it less slippery, or applying a hardening compound to further toughen the wearing surface. When a separate material—such as tile or a separate concrete fill—is to be applied as the wearing surface, the surface is usually kept deliberately rough. This may be achieved by simply reducing the degree of finished troweling.

Weather Exposure. Once the building is enclosed, interior floor slabs are not ordinarily exposed to exterior weather conditions. In cold climates, however, freezing and extreme temperature ranges should be considered if slabs are exposed to the weather. This may indicate the need for more temperature reinforcing, less distance between control joints, or the use of materials added to the concrete mix to enhance resistance to freezing.

6.2 Framed Floors on Grade

It is sometimes necessary to provide a concrete floor poured directly on the ground in a situation that precludes the use of a simple paving slab and requires a real structural spanning capability of the floor structure. One of these situations is where a deep foundation is provided for support of walls and columns and the potential settlement of upper ground masses may result in a breaking up and subsidence of the paving. Another situation is where considerable fill must be placed beneath the floor, and it is not feasible to produce a compaction of this amount of fill to assure a steady support for the floor.

Figure 6.1 illustrates two techniques that may be used to provide what amounts to a framed concrete slab and beam system poured directly on the ground. Where spans are modest and beam sizes not excessive it may be possible to produce the system in a single pour by

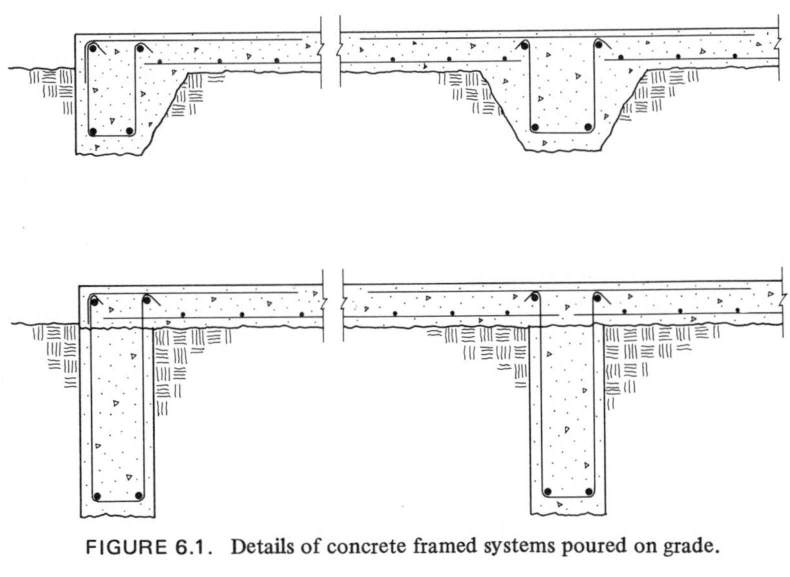

FIGURE 6.1. Details of concrete framed systems poured on grade.

simply trenching for the beam forms, as shown in the upper illustration in the figure. When large beams must be provided, it may be more feasible to use the system shown in the lower part of the figure. In this case the stems of the large beams are formed and poured and the slab is poured separately on the fill placed between the beam stems. These two techniques can be blended, of course, with smaller beams trenched in the fill between the large formed beams.

If the system with separately poured beams and slabs is used, it is necessary to provide for the development of shear between the slab and the top of the beam stems. Depending on the actual magnitude of the shear stresses involved, this may be done by various means. If stress is low, it may be sufficient to require a roughening of the surface of the top of the beam stems. If stress is of significant magnitude, shear keys, similar to those used for shear walls, may be used. If stirrups or ties are used in the beam stems, these will extend across the joint and assist in the development of shear.

6.3 Tension Anchors

Tension-resistive foundations are a special, although not unique, problem. Some of the situations that require this type of foundation are the following:

Anchorage of very light weight structures, such as tents, air-inflated structures, light metal buildings, and so on.

Anchorage of cables for tension structures or for guyed towers.

Anchorage for uplift resistance as part of the development of overturn resistance for the lateral bracing system for a building.

Figure 6.2 illustrates a number of elements that may be used for tension anchorage. The simple tent stake is probably the most widely used temporary tension anchor. It has been used in sizes ranging from large nails up to the huge stakes used for large circus tents. Also commonly used is the screw-ended stake, which offers the advantages of being somewhat more easily inserted and withdrawn, and having less tendency to loosen.

Ordinary concrete bearing foundations offer resistance to tension in the form of their own dead weight. The so-called dead man anchor

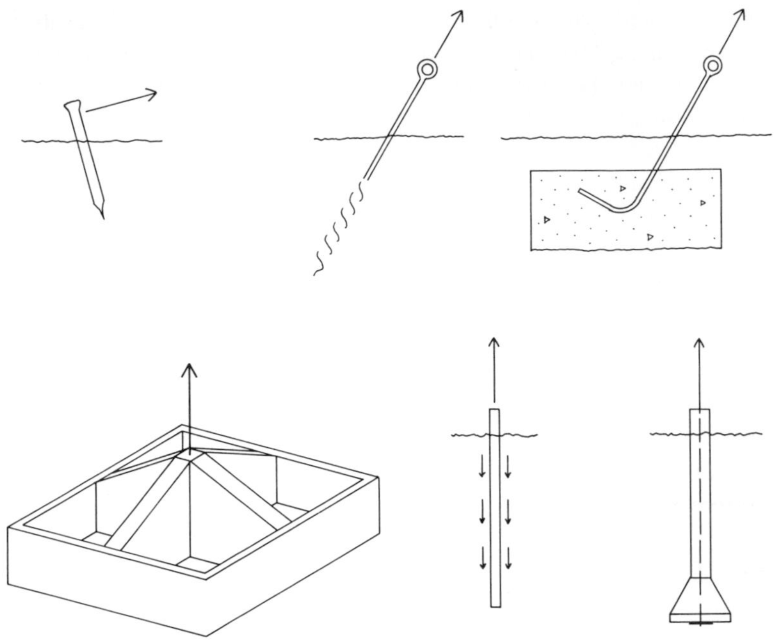

FIGURE 6.2. Various forms of tension anchors.

consists simply of a buried block of concrete, similar to a simple footing. Column and wall footings, foundation walls, concrete and masonry piers, and other such heavy elements may be utilized for this type of anchorage. Many lightweight building structures are essentially anchored by being fastened to their heavy foundations.

Where resistance to exceptionally high uplift force is required, special anchoring foundations may be used that develop resistance through a combination of their own dead weight plus the ballast effect of earth fill placed in or on them. Friction piles and piers may also be used for major uplift resistance, although if their shafts are of concrete, care should be taken to reinforce them adequately for the tension force. A special technique is to use a belled pier to resist force by its own dead weight plus that of the soil above it, since the soil must be pushed up by the bell in order to extract the pier. One method for the development of the tension force through the bell end is to anchor a cable to a large plate, which is cast into the bottom of the bell, as shown in Figure 6.2.

The nature of tension forces must be considered as well as their magnitude. Forces caused by wind or seismic shock will have a jarring effect that can loosen or progressively weaken anchorage elements. If the surrounding soil is soft and easily compressed, the effectiveness of the anchor may be reduced.

6.4 Temporary Foundations

Temporary structures of significant size offer a special foundation problem, since the usual buried foundation structure is not generally desired. If the period of use of the structure is quite short it may be possible to develop a structure for a foundation that simply rests on the top of the ground–a situation not acceptable for permanent structures. Thus an ordinary cast concrete footing may be poured on top of the ground and relatively easily removed when no longer required. If intended for removal and reuse, the footing may consist of a preformed element, such as that shown in Figure 6.3. Since the loading of such an element will usually cause some significant settlement, it may be advisable to place a cushion of compacted gravel beneath such an element for a more positive base. However, if the loads are relatively light, it may be acceptable to merely let the element seat itself by compressing the ground surface.

Figure 6.4 illustrates the use of a footing consisting of a grillage of treated timbers. This foundation offers the advantage of being somewhat easier to handle than a large concrete element as well as being easier to remove when no longer required.

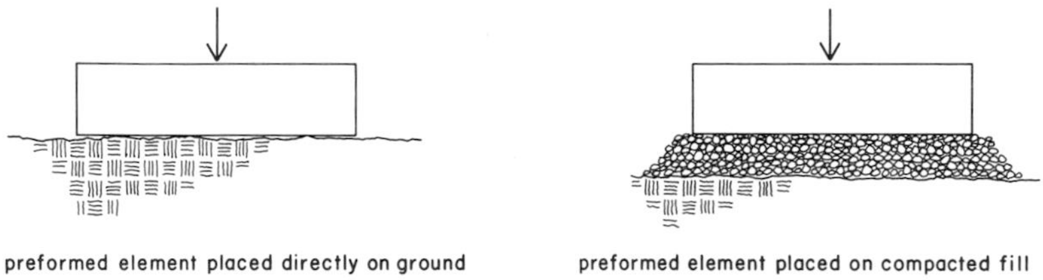

preformed element placed directly on ground
preformed element placed on compacted fill

FIGURE 6.3. Use of preformed foundation elements.

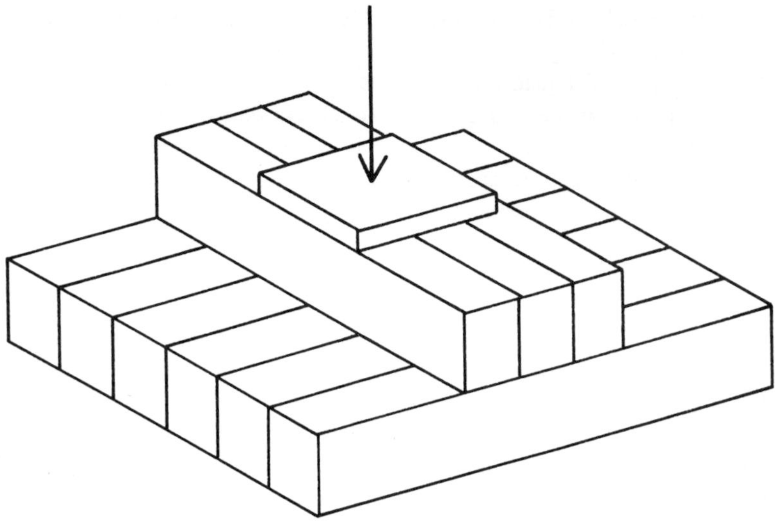

FIGURE 6.4. Temporary column foundation consisting of crossed timbers.

6.5 Hillside Foundations

Structures built on sloping sites offer a number of potential problems. Figure 6.5 illustrates a basic consideration for what is called the "daylight" dimension, which is the horizontal distance from the bottom of the footing to the adjacent ground surface. Many building codes require a minimum distance for this dimension in order to assure some safety against the pushing out of the footing in the down-slope direction.

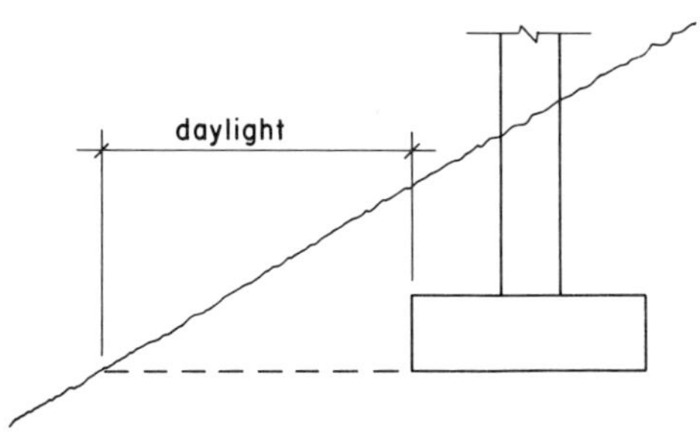

FIGURE 6.5. Critical exposure condition for a hillside footing.

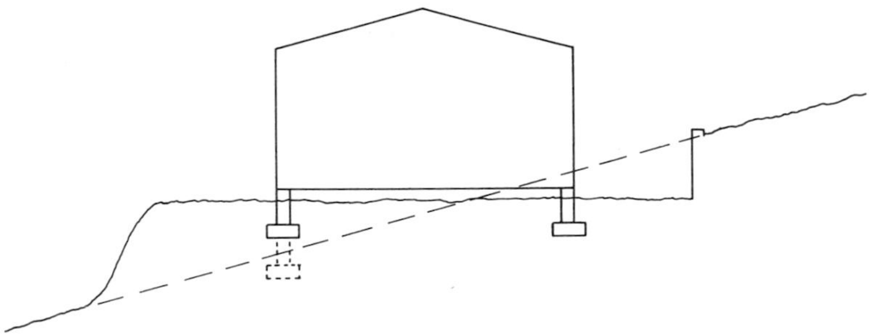

FIGURE 6.6. Problems of foundations on graded hillside sites.

Logically the limit for this distance should depend on the type of soil and the angle of the slope of the ground surface. For low slope angles a reasonable daylight dimension is assured simply because of the usual requirements for a minimum depth of the bottom of the footing below the surface in a vertical direction. As the slope increases, the daylight dimension is less assured and should be limited to some minimum distance.

A common problem with hillside construction is that shown in Figure 6.6, where recontouring of the construction site results in a portion of the building being placed on some significant depth of fill. If possible, all of the footings should be carried down to a depth below the fill. Where this is not feasible it may be necessary to use deep foundation elements of piles or piers. If the structure is relatively light and the fill is of proper materials placed with reasonable compaction, it may be possible to rest footings. in the fill, although such work should be done only with consultation and inspection by an experienced soils engineer.

References

For basic soil mechanics and foundation engineering:

1. R. B. Peck, W. E. Hanson, and T. H. Thornburn, *Foundation Engineering*, 2nd ed., Wiley, New York, 1974.
2. G. B. Sowers and G. F. Sowers, *Introductory Soil Mechanics and Foundations*, 3rd ed., Macmillan, New York, 1970.

For design of reinforced concrete:

3. Harry Parker, *Simplified Design of Reinforced Concrete*, 4th ed. (prepared by Harold D. Hauf), Wiley, New York, 1976.
4. Phil Ferguson, *Reinforced Concrete Fundamentals*, 3rd ed., Wiley, New York, 1974.
5. P. F. Rice and E. S. Hoffman, *Structural Design Guide to the ACI Building Code*, 2nd ed., Van Nostrand Reinhold, New York, 1979.
6. *CRSI Handbook*, 3rd ed., Concrete Reinforcing Steel Institute, 1978. (Based on strength design.)
7. *Reinforced Concrete Design Handbook: Working Stress Method*, Publication SP-3, 3rd ed., American Concrete Institute, 1965.

For elementary structural analysis:

8. Harry Parker, *Simplified Mechanics and Strength of Materials*, 3rd ed. (prepared by Harold D. Hauf), Wiley, New York, 1977.

Additional references:

9. *Uniform Building Code*, 1979 ed., International Conference of Building Officials, 5360 South Workman Mill Road, Whittier, CA 90601.
10. *Building Code Requirements for Reinforced Concrete*, ACI 318-77, Ameri-

can Concrete Institute, Box 19150, Redford Station, Detroit, MI 48219, 1977. (Usually referred to simply as the *ACI Code*.)

11. *Building Code Requirements for Reinforced Concrete*, ACI 318-63, American Concrete Institute, 1963. (This was the last edition that fully documented the working stress method; portions of this edition are reproduced in the appendix of this book.)
12. Charles G. Ramsey and Harold R. Sleeper, *Architectural Graphic Standards*, 8th ed., Wiley, New York, 1980.
13. James Ambrose and Dimitry Vergun, *Simplified Building Design for Wind and Earthquake Forces*, Wiley, New York, 1980.
14. *Masonry Design Manual*, 3rd ed., Masonry Institute of America, 2550 Beverly Boulevard, Los Angeles, CA 90057, 1979.
15. *Manual of Standard Practice for Detailing Reinforced Concrete Structures*, ACI 315, American Concrete Institute. (Commonly referred to as the *ACI Detailing Manual*.)

Glossary

The material presented here consists of a dictionary of the major words and terms from the field of foundation design that have been used in the work in this book. For a fuller explanation of most entries the reader should use the Index to find the related discussion in the text.

Active lateral pressure. See *Lateral pressure*.

Adobe. Masonry construction that utilizes unburned clay units.

Allowable stress. See *Stress*.

Anchorage. Refers to attachment for resistance to movement, usually a result of uplift, overturn, sliding, or horizontal separation.

Angle of internal friction (ϕ). Property that indicates shear strength of a cohesionless soil.

Atterberg limits. A number of properties and relationships that relate to the identification of cohesive soils. The major tested values are the water contents indicated by the following:

Liquid limit (w_L) is the limit for water content above which the soil displays the character of a liquid with the solid particles carried in suspension.

Plastic limit (w_P) is the water content at the boundary between the physical states of plastic (easily moldable) and solid (not moldable without fracture).

Shrinkage limit (w_s) is the water content at which the soil mass attains its least volume upon drying out.

The numeric difference between the liquid and plastic limits is called the *plasticity index* (I_p) and the values for the liquid limit and the plasticity index are displayed on the *plasticity chart*, which is used to classify the soil as a clay or silt. Soils with high values for both liquid limit and plasticity index are called *fat*; those with low values for both are called *lean*.

Backfill. See *Fill*.

Basement. Enclosed space within a building that is partly or wholly below the level of the ground surface.

Bearing foundation. Foundation that transfers loads to soil by direct vertical contact pressure. Usually refers to a *shallow bearing foundation*, which is a foundation that is placed directly beneath the lowest part of the building. See also *Footing*.

Bedrock. The level at which a large rock mass is considered to be solid, stable, and strong. Is typically overlain by masses that are fractured, weathered, or otherwise less stable and strong.

Caisson. See *Pier*.

Clay. General name for soil of very fine grain size and highly cohesive nature. Change in water content dramatically affects consistency, causing change from soft (easily remoldable) to hard (rocklike). *Fat* clays are those with high values for both liquid limit and plasticity index; *lean* clays are those with low values. Major structural property is unconfined compressive strength (q_u).

Coarse fraction. The portion of the solid particles of a soil sample with a grain size larger than the No. 200 sieve (0.003 in. or 0.075 mm). Generally consists primarily of sand and gravel.

Cohesionless. General lack of cohesiveness; noncohesive. The typical character of clean sands and gravels containing a small amount of fines.

Cohesive. General character of a soil in which the soil particles adhere to each other to produce a nondisintegrating mass. The typical character of fine-grained soils: silts and clays.

Column. A linear compression member. See also *Pier*.

Compaction. Action that tends to lower the void ratio and increase the density of a soil mass. When produced by artificial means, the degree of compaction obtained is measured in percent with reference to the theoretical minimum volume of the soil.

Compressibility. The relative resistance of a soil mass to volume change upon being subjected to compressive stress.

Consistency. The property of a cohesive soil that generally describes its physical state, ranging from soft to hard.

Consolidation. Volume reduction in a soil mass produced by a lowering of the void ratio. The effect resulting from compaction, shrinkage, and so on.

Crawl space. Space between the underside of the floor construction and the ground surface that occurs when a framed floor is suspended above the ground but there is no basement.

Curb. An edging strip, often occurring at the edge of a pavement; sometimes effects a small change in surface elevations on opposite sides of the curb.

Cut. Refers to the removal of existing soil deposits during the recontouring (or grading) of the ground surface. See *Grading*.

Dead Load. See *Load*.

Deep foundation. Foundation system that utilizes elements to achieve a considerable extension of the construction below the level of the bottom of the supported structure. Elements most commonly used are *piles* or *piers*.

Density. The weight per unit volume of a physical substance.

End-bearing pile. See *Pile*.

Equalized settlement. The design method in which a set of foundations is designed for equal settlement under dead load, rather than for a uniform bearing pressure under total load.

Erosion. Progressive removal of a soil mass due to water, wind, or other effects.

Excavation. Removal of soil mass to permit construction.

Expansive soil. Soil that has a tendency to increase its volume because of change in its water content.

Exploration. The general activity undertaken to identify and classify the constituent elements of a soil mass.

Fat clay. See *Clay*.

Fill. Usually refers to a soil deposit produced by other than natural effects. *Backfill* is soil deposited in the excessive part of an excavation after completion of the construction.

Fines. The portion of the solid particles of a soil sample that have a grain size smaller than the No. 200 sieve (0.003 in. or 0.075 mm). See also *Coarse fraction*.

Flocculent soil structure. Soil structure with a high void due to the bonding of soil particles into numerous, small cavelike cells. May be stable and strong under static conditions, but will often experience major volume reduction when subjected to saturation or shock.

Footing. A shallow, bearing-type foundation element, consisting typically of concrete that is poured directly into an excavation.

Foundation. The element, or system of elements, that effects the transition between a supported structure and the ground.

Freestanding wall. See *Wall*.

Friction. Resistance to sliding developed at the contact face between two surfaces.

Friction pile. See *Pile*.

Frost. The action, or the result, of the freezing of water; usually as condensed water vapor on a surface, or as moisture within a porous material (such as the ground).

Frost heave. See *Heave*.

Frost line. The level to which freezing of water extends below the ground surface.

Gap-graded soil. See *Grading*.

Grade. The level of the ground surface. Usually refers to the *finished grade*, which is the recontoured surface after the completion of construction.

Grade beam. A horizontal element in a foundation system that serves some spanning or load-distributing function.

Grading. Has two usages.

1. Refers to the gradation of size of the solid particles in a sample of soil, which is qualified as *uniform* (major grouping within a limited size range), *well-graded* (wide range of sizes with no gaps), or *gap-graded* (wide range of sizes with an absence of particles of some sizes). Both uniform and gap-graded soils are considered *poorly graded*.
2. Refers to the general activity of recontouring the ground surface.

Gravel. Soil particles consisting of rocks and rock fragments of a size range from $\frac{3}{16}$ in. to 3 in. in diameter.

Grout. Lean concrete (predominantly water, cement and sand) used as a filler in the voids of hollow masonry units, under column base plates, and so on.

Grouted masonry. Masonry of hollow units in which the voids are filled with grout.

Grain. A discrete particle of the material that constitutes the solid portion of a soil mass.

Hardpan. A highly consolidated soil mass, commonly overlying a rock formation at some distance below the ground surface. Usually consists of a silt and clay mixture and has high bearing resistance and low compressibility.

Heave. Upward swelling action of a soil mass at the ground surface.

Kern. Limiting dimension for the eccentricity of a compression force, if tension stress is to be avoided.

Key. A slot or pocket formed at the construction joint between two concrete elements (such as between the bottom of a wall and its footing) for the resistance of shear action parallel to the joint face.

Lateral pressure. Horizontal soil pressure of two kinds:

1. *Active* lateral pressure is that exerted by a retained soil upon the retaining structure.
2. *Passive* lateral pressure is that exerted by soil against an object that is attempting to move in a horizontal direction.

Lean clay. See *Clay*.

Liquid limit (w_L). See *Atterberg limits*.

Live load. See *Load*.

Load. The active force (or combination of forces) exerted upon a structure. *Dead load* is permanent load due to gravity, which includes the weight of the structure itself. *Live load* is any load component that is not permanent, including those due to wind, seismic effects, temperature effects, and gravity forces that are not permanent.

Mandrel. In pile driving, refers to the use of a heavy steel filler element that is used to drive a light sheet steel liner into the ground; once the liner is in place, the mandrel is withdrawn and the liner is filled with concrete.

Mat foundation. A very large bearing-type foundation. When the entire bottom of a building is constituted as a single mat, it is also called a *raft foundation*.

Maximum density. The theoretical density of a soil mass achieved when the void is reduced to the minimum possible.

Mud-jacking. A technique for filling an eroded void or for raising a settled structure consisting of pumping grout or a soil-cement mixture under pressure beneath the structure to "jack" it up.

Noncohesive. See *Cohesionless*.

Organic. Refers to material of biological (plant or animal) origin.

Overturning effect. The toppling, tipping over, effect on a structure produced by lateral (horizontal) loads. Is commonly partly or wholly resisted by the dead load on the structure that produces the so-called *restoring moment*.

Passive lateral pressure. See *Lateral pressure*.

Pedestal. A short pier, or upright compression member. Is actually a short column with a ratio of unsupported height to least lateral dimension of three or less.

Penetration resistance (*N*). A standard, field-tested property with significance for cohesionless soils, consisting of the number of blows required to drive a standard sampling device into the soil mass. Generally serves as a direct index of the density and the compressibility of the soil.

Permeability. A measurement of the rate at which water will seep into, or drain out of, a soil mass.

Pier. Generally refers to a compression element, typically of rather stout proportions (versus slender). Also used to describe a deep foundation element that is placed in an excavation, rather than being driven as a pile. Although it actually refers to a particular method of excavation, the term *caisson* is also commonly used to describe a pier foundation.

Pile. A deep foundation element, consisting of a linear, shaftlike member, that is placed by being driven dynamically into the ground. *Friction piles* develop resistance to both downward load and upward (pull-out) load through friction between the soil and the surface of the pile shaft. *End-bearing piles* are driven so that their ends are seated in low-lying strata of rock or very hard soil.

Pile cap. The concrete element that is used to transfer load to a group of piles; it functions in a manner similar to a column footing.

Plastic clay. See *Clay*.

Plastic limit (w_p). See *Atterberg limits*.

Plasticity index. See *Atterberg limits*.

Poorly graded soil. See *Grading*.

Porosity (n). The percentage of void in a soil mass.

Preconsolidation. The condition of a highly compressed soil, usually used to refer to a condition produced by other than natural causes, as when the piling up of soil on the ground surface, vibration, or introduction of water is used to produce the consolidated state.

Presumptive bearing pressure. A value for allowable vertical bearing pressure that is permitted to be used in the absence of extensive investigation and testing. Requires a minimum of identification (loose, silty sand, etc.) and is usually quite conservative.

Quicksand. A soil deposit that is reasonably stable if undisturbed, but becomes suddenly quite loose and fluidlike when disturbed.

Raft foundation. See *Mat foundation*.

Restoring moment. See *Overturning effect*.

Retaining wall. A structure used to brace a vertical cut, or a change in elevation of the ground surface. The term is usually used to refer to a *cantilever retaining wall*, which is a freestanding structure consisting only of a wall and its footing, although basement walls also serve a retaining function.

Rock. See *Soil.*

Rock flour. A very fine grained material produced by the mechanical disintegration of rock; a byproduct of the grinding, crushing, or blasting of rock. It is actually the same material as sand or gravel, but has a particle size range in the category of silt.

Runoff. Flowing surface water, usually from rain or melting snow or ice. A major source of erosion effect on the ground surface.

Saturation. The condition that exists when the void in the soil is completely filled with water. The *degree of saturation* (S_r) is the expression of the ratio of the volume of water in the soil to the volume of the void, determined in percent. A condition of *partial saturation* exists when the void is only partly filled with water. *Oversaturation*, or *supersaturation*, occurs when the soil contains water in excess of the normal volume of the void, which usually results in some flotation or suspension of the solid particles in the soil.

Settlement. The general downward movement of a foundation caused by the loads and the reactions of the supporting ground.

Shallow bearing foundation. See *Bearing foundation.*

Shoring. Bracing; usually refers to the bracing of a cut, a shoreline, or some structure that is in danger of collapse due to erosion, undercutting, unequal settlement, and so on.

Shrinkage limit. See *Atterberg limits.*

Silt. Fine-grained soil material in the size range between sand and clay. Typically possesses properties that are transitional between those of the classic cohesionless material (sand) and the classic cohesive material (plastic clay).

Slab. A horizontal, planar element of concrete. *Slab-on-grade* refers to a floor that is a concrete pavement, poured directly on the ground surface.

Soil. In foundation and soil work a general distinction is made between soil and rock. *Soil* is any material that can be reduced to discrete particles or to a semifluid mass by the actions of water and minor agitation. *Rock* is any material that is generally not affected in its structural properties by variation in water content and that offers considerable resistance to excavation or to separation into discrete particles.

Specific gravity. A relative indication of density, using the weight of water as a reference. A specific gravity of 2 indicates a density (weight) of twice that of water (usually assumed to weigh 62.4 lb/ft^3.)

Stability. Refers to the inherent capability of something to maintain its shape, its physical state, or its ability to resist force. With reference to soil, usually refers to the relative capability of the soil to maintain its structural integrity against effects of time, temperature, frost, change in water content, shock, vibration, and so on.

Strata. Plural of stratum. Levels or layers. Used to describe a soil mass of singular character, typically existing in a layer between other soil masses of different character.

Stratum. See *Strata*.

Strength design. One of the two fundamental design techniques for assuring a margin of safety for a structure. *Stress design*, also called *working stress design*, is performed by analyzing stresses produced by the estimated actual usage loads, and assigning limits for the stresses that are below the ultimate capacity of the materials by some margin. *Strength design*, also called *ultimate strength design*, is performed by multiplying the actual loads by the desired factor of safety (the universal average factor being two) and proceeding to design a structure that will have that load as its ultimate failure load.

Stress. The mechanism of force within the material of a structure; visualized as a pressure effect (tension or compression) or a shear effect on the surface of a unit of the material, and quantified in units of force per unit area. *Allowable*, *permissible*, or *working* stress refers to a stress limit that is used in stress design methods. *Ultimate* stress refers to the maximum stress that is developed just prior to failure of the material.

Stress design. See *Strength design*.

Subsidence. Settlement of a soil mass, usually manifested by the sinking of the ground surface.

Surcharge. Vertical load applied at the ground surface or simply above the level of the bottom of a footing. The weight of soil above the bottom of the footing is surcharge for the footing.

Terrace. A flat shelf, or plateau, occurring in a generally sloping ground surface.

Ultimate strength. See *Strength design*.

Unconfined compressive strength (q_u). See *Clay*.

Underpinning. Propping up of a structure that is in danger of, or has already experienced, some support failure, notably due to excessive settlement, undermining, erosion, and so on.

Upheaval. Pushing upward of a soil mass.

Viscosity. The general measurement of the mobility, or flowing character, of a fluid or a semifluid mass. Heavy oil has high viscosity (is viscous); water has low viscosity.

Void. The portion of the volume of a soil that is not occupied by solid material; it is ordinarily filled partly with water and partly with air.

Void ratio (e). The term most commonly used to indicate the amount of void in a soil mass, expressed as the ratio of the volume of the void to the volume of the solids. See also *porosity*, which is the expression of the volume of the void as a percentage of the total volume of the soil.

Wall. A vertical, planar building element. *Foundation* walls are those that are partly or totally below ground. *Bearing* walls are used to carry vertical loads in direct compression. *Grade* walls are those that are used to achieve the transition between the building that is above the ground and the foundations that are below it, grade being used to refer to the level of the ground surface at the edge of the building. (See also *Grade beam*.) *Shear* walls are those used to brace the building against horizontal forces due to wind or seismic shock. *Freestanding* walls are walls whose tops are not laterally braced. *Retaining* walls are walls that resist horizontal soil pressure.

Working stress. See *Stress*.

Working stress design. See *Strength design*.

Appendix
Reinforced Concrete Analysis and Design—Working Stress Method

The following is a brief presentation of the formulas and procedures used in the working stress method.

A.1 Flexure

This discussion refers to a rectangular concrete section, tension reinforcing only.

Referring to Figure A.1, the following are defined:

b = the width of the concrete compression zone.

d = the effective depth of the section for stress analysis; from the centroid of the steel to the edge of the compression zone.

A_s = the cross sectional area of the reinforcing.

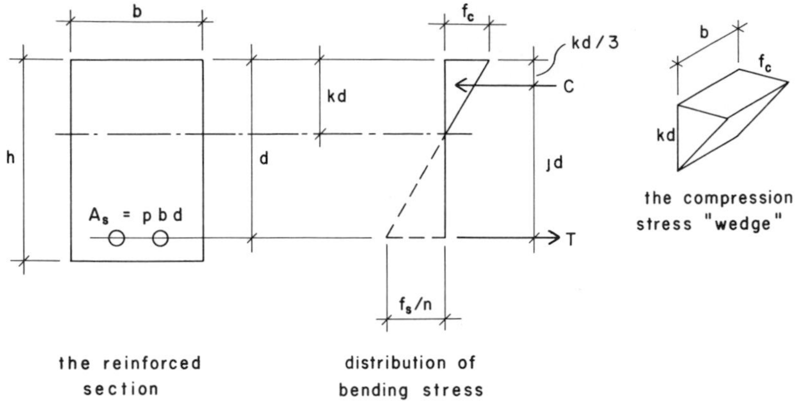

FIGURE A.1. Moment resistance of a rectangular concrete section with tension reinforcing.

p = the percentage of reinforcing, defined as

$$p = \frac{A_s}{bd}$$

$$n = \text{the elastic ratio} = \frac{E \text{ of the steel reinforcing}}{E \text{ of the concrete}}$$

kd = the height of the compression stress zone; used to locate the neutral axis of the stressed section; expressed as a percentage (k) of d.

jd = the internal moment arm, between the net tension force and the net compression force; expressed as a percentage (j) of d.

f_c = the maximum compressive stress in the concrete.

f_s = the tensile stress in the reinforcing.

The compression force, C, may be expressed as the volume of the compression stress "wedge," as shown in the figure.

$$C = \tfrac{1}{2}\,(kd)\,(b)\,(f_c) = \tfrac{1}{2}\,kf_c bd$$

Using the compression force the moment resistance of the section may be expressed as

$$M = Cjd = (\tfrac{1}{2}\,kf_c bd)\,(jd) = \tfrac{1}{2}\,kjf_c bd^2 \qquad (1)$$

This may be used to derive an expression for the concrete stress:

$$f_c = \frac{2M}{kjbd^2} \tag{2}$$

The resisting moment may also be expressed in terms of the steel and the steel stress as

$$M = Tjd = (A_s)(f_s)(jd)$$

This may be used for determination of the steel stress or for finding the required area of steel.

$$f_s = \frac{M}{A_s jd} \tag{3}$$

$$A_s = \frac{M}{f_s jd} \tag{4}$$

A useful reference is the so-called balanced section, which occurs when the exact amount of reinforcing used results in the simultaneous limiting stresses in the concrete and steel. The properties which establish this relationship may be expressed as follows:

$$\text{balanced } k = \frac{1}{1 + f_s/nf_c} \tag{5}$$

$$j = 1 - \frac{k}{3} \tag{6}$$

$$p = \frac{f_c k}{2f_s} \tag{7}$$

$$M = Rbd^2 \tag{8}$$

in which

$$R = \tfrac{1}{2}\, kjf_c \tag{9}$$

(derived from formula 1). If the limiting compression stress in the concrete ($f_c = 0.45 f_c'$) and the limiting stress in the steel are entered in Formula 5, the balanced section value for k may be found. Then the corresponding values for j, p and R may be found. The balanced p may be used to determine the maximum amount of tensile reinforcing that may be used in a section without the addition of compressive reinforcing. If less tensile reinforcing is used, the moment will be limited by the steel stress, the maximum stress in the concrete will be below the limit of 0.45 f_c', the value of k will be slightly lower than the balanced value and the value of j slightly higher than the balanced value. These relationships are useful in design for the determination of approximate requirements for cross sections.

Table A.1 gives the balanced section properties for various combinations of concrete strength and limiting steel stress. The values of n, k, j, and p are all without units. However, R must be expressed in particular units; the unit used in the table is kip-inches (k-in.).

When the area of steel used is less than the balanced p, the true value of k may be determined by the following formula:

$$k = \sqrt{2np - (np)^2} - np \tag{10}$$

Figure A.2 may be used to find approximate k values for various combinations of p and n.

TABLE A.1 Balanced Section Properties for Rectangular Concrete Sections With Tension Reinforcing Only

f_s (psi)	f_c' (psi)	n	k	j	p	R (k-in.)
16,000	2000	11.3	0.389	0.870	0.0109	152
	2500	10.1	0.415	0.862	0.0146	201
	3000	9.2	0.437	0.854	0.0184	252
	4000	8.0	0.474	0.842	0.0266	359
20,000	2000	11.3	0.337	0.888	0.0076	135
	2500	10.1	0.362	0.879	0.0102	179
	3000	9.2	0.383	0.872	0.0129	226
	4000	8.0	0.419	0.860	0.0188	324
24,000	2000	11.3	0.298	0.901	0.0056	121
	2500	10.1	0.321	0.893	0.0075	161
	3000	9.2	0.341	0.886	0.0096	204
	4000	8.0	0.375	0.875	0.0141	295

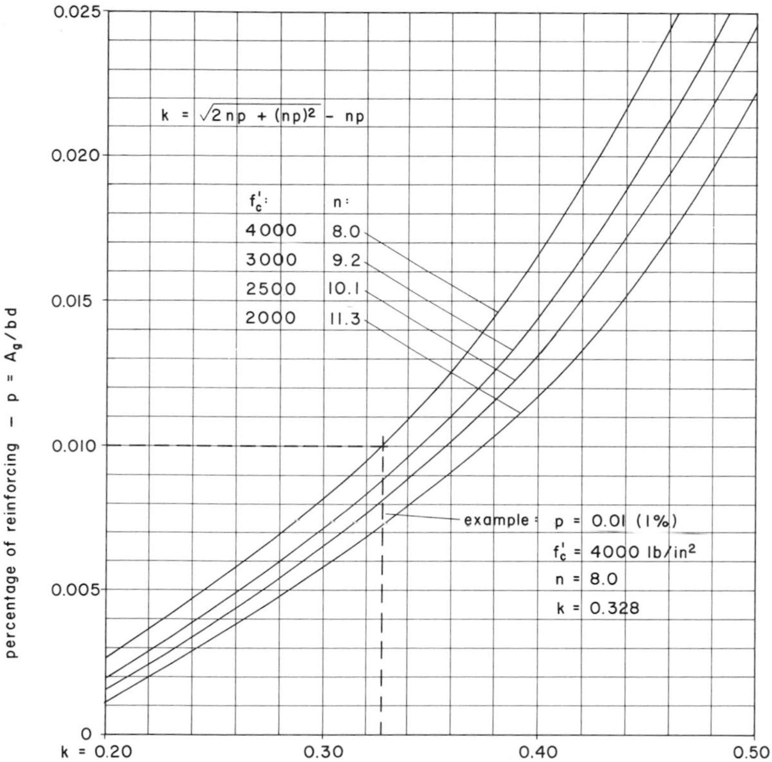

FIGURE A.2. *k* factors for rectangular concrete sections with tension reinforcing–as a function of *p* and *n*.

A.2 Shear

Three shear stress situations must be considered, as shown in Figure A.3. The first of these is so-called pure or direct shear as developed on keys, brackets and short cantilevers. The shear stress in this case is calculated as the shear force divided by the gross concrete cross section if no reinforcing is provided. The codes generally do not provide any guides for this situation, so designers must use their own judgment. The author recommends the use of a maximum stress of 10% of f_c' for such situations, limited to the condition that the length of the canti-

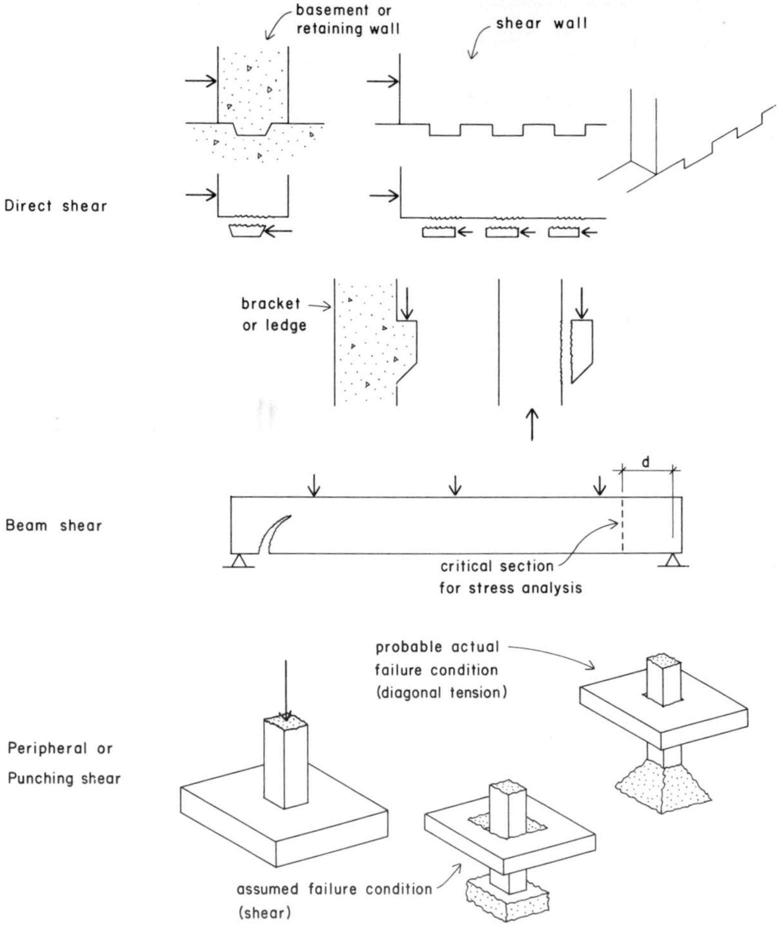

FIGURE A.3. Shear actions in concrete structures.

lever is less than the thickness of the section in the direction of the shear force.

The second condition for shear is that which occurs in beams. In this case the critical maximum stress is determined at a distance from the support equal to the effective depth d. The concrete alone is permitted to take a stress of $1.1\sqrt{f_c'}$ with any additional stress required to be developed with shear reinforcing.

The third case is that which occurs as so-called punching shear. Two examples of this are the shear around columns in flat slab systems and the shear around columns in isolated footings. The code refers to this as peripheral shear. The critical section is defined by circumscribing the column at a distance of one half the effective depth d, and the total shear force (the column load) is assumed to produce a uniform stress on a section consisting of the product of this circumference times the effective depth. In this case the limiting stress is $2\sqrt{f_c'}$.

In summary, the three shear conditions are treated as follows:

1. Direct shear

$$v_c = \frac{V}{bh}, \text{ where } bh \text{ is the gross area of concrete}$$

maximum stress = $0.10 f_c'$

2. Beam shear

$$v_c = \frac{V}{bd}, \text{ where } V \text{ is taken at } d \text{ distance from supports}$$

maximum stress = $1.1\sqrt{f_c'}$

3. Peripheral shear

$$v_c = \frac{V}{\Sigma wd}, \text{ where } \Sigma w \text{ is the peripheral circumference}$$

maximum stress = $2\sqrt{f_c'}$

A.3 Bond and Anchorage

Work in this book involves the following situations of bond and anchorage:

1. ***Bond stress on reinforcing for flexural tension.*** In this case the stress is calculated as follows:

$$u = \frac{V}{\Sigma_0 jd}$$

in which V is the flexural shear and Σ_0 is the sum of the perimeters of the bars.

Stress is limited to the following:

$$u = \frac{3.4\sqrt{f_c'}}{D} \text{ or a maximum of 350 psi for top bars}$$

in which D is the diameter of the bars.

Top bars are any horizontal bars with more than 12 in. of concrete below them.

$$u = \frac{4.8\sqrt{f_c'}}{D} \text{ or a maximum of 500 psi for other bars.}$$

Allowable bond stresses based on these requirements are given for bar sizes 3 through 11 in Table A.2.

2. ***Tension anchorage, or pullout resistance.*** This may be developed by bond stress in embedment, by hooking the bar end, or by anchoring the end of the bar with some mechanical device.

   The allowable stress for bond is the same as for "other" bars, as given above.

   A so-called standard hook must conform to requirements stipulated in the code. Hooks are limited to the development of a maximum of 10,000 psi in the bars, thus the full development of bars requires some embedment length in addition to the hook.

   End anchorage by mechanical means is sometimes necessary where details of the construction preclude a hook or sufficient length for embedment. These situations seldom occur in foundations.
3. ***Compression anchorage.*** In foundations compression anchorage occurs primarily in the development of doweling for the vertical reinforcement in columns or piers. It is recommended that reinforced concrete column design in general, including design for anchorage, be done with the latest methods of strength design. No design of this situation is given in this book. For short piers with minimum reinforcing, as discussed in Section 3.5 of this book, dowels may be designed with the following criteria:

$$u = \frac{6.5\sqrt{f_c'}}{D}, \text{ or a maximum of 400 psi}$$

TABLE A.2. Allowable Bond Stress for Flexural Members with A305 Deformations

f'_c (psi)	Bar Numbers								
	3	4	5	6	7	8	9	10	11
Top bars: $u = \frac{3.4\sqrt{f'_c}}{D}$ or maximum of 350 psi									
2000	350	304	243	203	174	152	135	120	108
3000	350	350	298	248	213	186	165	147	132
4000	350	350	344	287	246	215	191	169	152
Other than top bars: $u = \frac{4.8\sqrt{f'_c}}{D}$ or maximum of 500 psi									
2000	500	429	343	286	245	215	190	169	152
3000	500	500	420	350	300	263	233	207	186
4000	500	500	486	465	347	304	269	239	215

A.4 Compression Members

Design of compression members is limited to short axially loaded piers with minimum reinforcing in this book. For these the following is recommended.

1. Compression design should be done using whichever of the following is critical.
   a. Consideration of the pier as a pure, unreinforced compression element, using the allowable stress for a member with bearing on the full area

   $$f_c = 0.25\, f_c'$$

   b. Consideration of the pier as an axially loaded tied column with reinforcing limited to the minimum percentage of 0.5% of the gross concrete section. For this the maximum allowable total load is determined as

   $$P = A_g(0.25\, f_c' + f_s p_g)(0.85)$$

   in which f_s is limited to 40% of f_y.
2. Development of the doweling of the pier reinforcement may be done as described in the preceeding discussion on anchorage.
3. Development of any dowels from concrete columns on top of the pier should be done with strength design methods. This may be critical in establishing the minimum height of the pier.
4. Design for bearing stress under columns resting on top of piers or directly on top of footings. For piers the critical relationship is between the area of bearing contact and the cross sectional area of the pier. The allowable stress in bearing is bracketed between two limiting conditions, as follows:

   $$f_c = 0.25\, f_c'$$

   when the bearing is on the full area of the pier.

   $$f_c = 0.375\, f_c'$$

   when the bearing is on one third or less of the area of the pier cross section. For conditions falling between these limits the allowable stress may be proportioned linearly. For footings the bearing area will usually be less than one third of the footing area, so the higher stress limit usually prevails.

A.5 Excerpts from the 1963 *ACI Code*

The material on the following pages consists of selected reprints from the 1963 ACI Code: *Building Code Requirements for Reinforced Concrete*, ACI 318-63 (Ref. 11). This material is reprinted with permission of the publishers, American Concrete Institute, Box 19150, Redford Station, Detroit, MI 48219. This is the principal reference for the structural analysis and design by the working stress method as it is utilized in the example calculations in this book.

TABLE 1002(a)—ALLOWABLE STRESSES IN CONCRETE

Description		Allowable stresses: For any strength of concrete in accordance with Section 502	For strength of concrete shown below: $f_c' =$ 2500 psi	$f_c' =$ 3000 psi	$f_c' =$ 4000 psi	$f_c' =$ 5000 psi
Modulus of elasticity ratio: n		$\frac{29,000,000}{w^{1.5}\,33\sqrt{f_c'}}$				
For concrete weighing 145 lb per cu ft (see Section 1102)	n		10	9	8	7
Flexure: f_c						
Extreme fiber stress in compression	f_c	$0.45f_c'$	1125	1350	1800	2250
Extreme fiber stress in tension in plain concrete footings and walls	f_c	$1.6\sqrt{f_c'}$	80	88	102	113
Shear: v (as a measure of diagonal tension at a distance d from the face of the support)						
Beams with no web reinforcement*	v_c	$1.1\sqrt{f_c'}$	55*	60*	70*	78*
Joists with no web reinforcement	v_c	$1.2\sqrt{f_c'}$	61	66	77	86
Members with vertical or inclined web reinforcement or properly combined bent bars and vertical stirrups	v	$5\sqrt{f_c'}$	250	274	316	354
Slabs and footings (peripheral shear, Section 1207)*	v_c	$2\sqrt{f_c'}$	100*	110*	126*	141*
Bearing: f_c						
On full area		$0.25f_c'$	625	750	1000	1250
On one-third area or less†		$0.375f_c'$	938	1125	1500	1875

*For shear values for lightweight aggregate concrete see Section 1208.
†This increase shall be permitted only when the least distance between the edges of the loaded and unloaded areas is a minimum of one-fourth of the parallel side dimension of the loaded area. The allowable bearing stress on a reasonably concentric area greater than one-third but less than the full area shall be interpolated between the values given.

1201—Shear stress*†

(a) The nominal shear stress, as a measure of diagonal tension, in reinforced concrete members shall be computed by:

$$v = V/bd \qquad (12\text{-}1)^*$$

For design, the maximum shear shall be considered as that at the section a distance, d, from the face of the support.‡ Wherever applicable, effects of torsion shall be added and effects of inclined flexural compression in variable-depth members shall be included.

*The term j is omitted in determination of nominal shear stress.
†Special provisions for lightweight aggregate concretes are given in Section 1208.
‡This provision does not apply to brackets and other short cantilevers.

1207—Shear stress in slabs and footings*

(a) The shear capacity of slabs and footings in the vicinity of concentrated loads or concentrated reactions shall be governed by the more severe of two conditions:

1. The slab or footing acting essentially as a wide beam, with a potential diagonal crack extending in a plane across the entire width. This case shall be considered in accordance with Section 1201.

2. Two-way action existing for the slab or footing, with potential diagonal cracking along the surface of a truncated cone or pyramid around the concentrated load or reaction. The slab or footing in this case shall be designed as required in the remainder of this section.

(b) The critical section for shear to be used as a measure of diagonal tension shall be perpendicular to the plane of the slab and located at a distance $d/2$ out from the periphery of the concentrated load or reaction area.

(c) The nominal shear stress shall be computed by:

$$v = V/b_o d \qquad (12\text{-}8)$$

in which V and b_o are taken at the critical section specified in (b). The shear stress, v, so computed shall not exceed $2\sqrt{f_c'}$, unless shear rein-

*For transfer of moments and effects of openings see Section 920.

CHAPTER 13 — BOND AND ANCHORAGE — WORKING STRESS DESIGN

1300—Notation

d = distance from extreme compression fiber to centroid of tension reinforcement
D = nominal diameter of bar, inches
f_c' = compressive strength of concrete (see Section 301)
j = ratio of distance between centroid of compression and centroid of tension to the depth, d
Σo = sum of perimeters of all effective bars crossing the section on the tension side if of uniform size; for mixed sizes, substitute $4A_s/D$, where A_s is the total steel area and D is the largest bar diameter. For bundled bars use the sum of the exposed portions of the perimeters
u = bond stress
V = total shear

1301—Computation of bond stress in flexural members

(a) In flexural members in which the tension reinforcement is parallel to the compression face, the flexural bond stress at any cross section shall be computed by

$$u = \frac{V}{\Sigma o j d} \qquad (13\text{-}1)$$

Bent-up bars that are not more than $d/3$ from the level of the main longitudinal reinforcement may be included. Critical sections occur at the face of the support, at each point where tension bars terminate within a span, and at the point of inflection.

(b) To prevent bond failure or splitting, the calculated tension or compression in any bar at any section must be developed on each side of that section by proper embedment length, end anchorage, or, for tension only, hooks. Anchorage or development bond stress, u, shall be computed as the bar forces divided by the product of Σo times the embedment length.

(c) The bond stress, u, computed as in (a) or (b) shall not exceed the limits given below, except that flexural bond stress need not be considered in compression, nor in those cases of tension where anchorage bond is less than 0.8 of the permissible.

(1) For tension bars with sizes and deformations conforming to ASTM A 305:

Top bars*	$\frac{3.4\sqrt{f_c'}}{D}$ nor 350 psi
Bars other than top bars	$\frac{4.8\sqrt{f_c'}}{D}$ nor 500 psi

*Top bars, in reference to bond, are horizontal bars so placed that more than 12 in. of concrete is cast in the member below the bar.

(2) For tension bars with sizes and deformations conforming to ASTM A 408:

Top bars*	$2.1\sqrt{f_c'}$
Bars other than top bars	$3\sqrt{f_c'}$

(3) For all deformed compression bars:

$$6.5\sqrt{f_c'} \text{ nor } 400 \text{ psi}$$

(4) For plain bars the allowable bond stresses shall be one-half of those permitted for bars conforming to ASTM A 305 but not more than 160 psi.

(d) Adequate anchorage shall be provided for the tension reinforcement in all flexural members to which Eq. (13-1) does not apply, such as sloped, stepped or tapered footings, brackets, or beams in which the tension reinforcement is not parallel to the compression face.

*Top bars, in reference to bond, are horizontal bars so placed that more than 12 in. of concrete is cast in the member below the bar.

CHAPTER 23 — FOOTINGS

2301—Scope

(a) The requirements prescribed in Sections 2302 through 2309 apply only to isolated footings.

(b) General procedures for the design of combined footings are given in Section 2310.

2302—Loads and reactions

(a) Footings shall be proportioned to sustain the applied loads and induced reactions without exceeding the stresses or strengths prescribed in Parts IV-A and IV-B, and as further provided in this chapter.

(b) In cases where the footing is concentrically loaded and the member being supported does not transmit any moment to the footing, computations for moments and shears shall be based on an upward reaction assumed to be uniformly distributed per unit area or per pile and a downward applied load assumed to be uniformly distributed over the area of the footing covered by the column, pedestal, wall, or metallic column base.

(c) In cases where the footing is eccentrically loaded and or the member being supported transmits a moment to the footing, proper allowance shall be made for any variation that may exist in the intensities of reaction and applied load consistent with the magnitude of the applied load and the amount of its actual or virtual eccentricity.

(d) In the case of footings on piles, computations for moments and shears may be based on the assumption that the reaction from any pile is concentrated at the center of the pile.

2303—Sloped or stepped footings

(a) In sloped or stepped footings, the angle of slope or depth and location of steps shall be such that the allowable stresses are not exceeded at any section.

(b) In sloped or stepped footings, the effective cross section in compression shall be limited by the area above the neutral plane.

(c) Sloped or stepped footings that are designed as a unit shall be cast as a unit.

2304—Bending moment

(a) The external moment on any section shall be determined by passing through the section a vertical plane which extends completely across the footing, and computing the moment of the forces acting over the entire area of the footing on one side of said plane.

(b) The greatest bending moment to be used in the design of an isolated footing shall be the moment computed in the manner prescribed in (a) at sections located as follows:

1. At the face of the column, pedestal or wall, for footings supporting a concrete column, pedestal or wall
2. Halfway between the middle and the edge of the wall, for footings under masonry walls
3. Halfway between the face of the column or pedestal and the edge of the metallic base, for footings under metallic bases

(c) The width resisting compression at any section shall be assumed as the entire width of the top of the footing at the section under consideration.

(d) In one-way reinforced footings, the total tensile reinforcement at any section shall provide a moment of resistance at least equal to the moment computed as prescribed in (a); and the reinforcement thus determined shall be distributed uniformly across the full width of the section.

(e) In two-way reinforced footings, the total tension reinforcement at any section shall provide a moment of resistance at least equal to the moment computed as prescribed in (a); and the total reinforcement thus determined shall be distributed across the corresponding resisting section as prescribed for square footings in (f), and for rectangular footings in (g).

(f) In two-way square footings, the reinforcement extending in each direction shall be distributed uniformly across the full width of the footing.

(g) In two-way rectangular footings, the reinforcement in the long direction shall be distributed uniformly across the full width of the footing. In the case of the reinforcement in the short direction, that portion determined by Eq. (23-1) shall be uniformly distributed across a band-width (B) centered with respect to the center line of the column or pedestal and having a width equal to the length of the short side of the footing. The remainder of the reinforcement shall be uniformly distributed in the outer portions of the footing.

$$\frac{\textit{Reinforcement in band-width } (B)}{\textit{Total reinforcement in short direction}} = \frac{2}{(S+1)} \quad \text{(23-1)}$$

where S is the ratio of the long side to the short side of the footing.

2305—Shear and bond

(a) For computation of shear in footings, see Section 1207 or 1707.

(b) Critical sections for bond shall be assumed at the same planes as those prescribed for bending moment in Section 2304(b)1; also at all other vertical planes where changes of section or of reinforcement occur.

(c) Computation for shear to be used as a measure of flexural bond shall be based on a vertical section which extends completely across the footing, and the shear shall be taken as the sum of all forces acting over the entire area of the footing on one side of such section.

(d) The total tensile reinforcement at any section shall provide a bond resistance at least equal to the bond requirement as computed from the external shear at the section.

(e) In computing the external shear on any section through a footing supported on piles, the entire reaction from any pile whose center is located 6 in. or more outside the section shall be assumed as producing shear on the section; the reaction from any pile whose center is located 6 in. or more inside the section shall be assumed as producing no shear on the section. For intermediate positions of the pile center, the portion of the pile reaction to be assumed as producing shear on the section shall be based on straight-line interpolation between full value at 6 in. outside the section and zero value at 6 in. inside the section.

(f) For allowable shearing values, see Sections 1207 and 1707.

(g) For allowable bond values, see Sections 1301 (c) and 1801 (c).

2306—Transfer of stress at base of column

(a) The stress in the longitudinal reinforcement of a column or pedestal shall be transferred to its supporting pedestal or footing either by extending the longitudinal bars into the supporting member, or by dowels.

(b) In case the transfer of stress in the reinforcement is accomplished by extension of the longitudinal bars, they shall extend into the supporting member the distance required to transfer this stress to the concrete by bond.

(c) In cases where dowels are used, their total sectional area shall be not less than the sectional area of the longitudinal reinforcement in the member from which the stress is being transferred. In no case shall the number of dowels per member be less than four and the diameter of the dowels shall not exceed the diameter of the column bars by more than 1/8 in.

(d) Dowels shall extend up into the column or pedestal a distance at least equal to that required for lap of longitudinal column bars [see Section 805] and down into the supporting pedestal or footing the distance required to transfer to the concrete, by allowable bond stress, the full working value of the dowel [see Section 918(i)].

(e) The compression stress in the concrete at the base of a column or pedestal shall be considered as being transferred by bearing to the top of the supporting pedestal or footing. The compression stress on the loaded area shall not exceed the bearing stress allowable for the quality of concrete in the supporting member as determined by the ratio of the loaded area to the supporting area.

(f) For allowable bearing stresses, design by Part IV-A shall conform to Table 1002(a), and for design by Part IV-B to 1.9 times those values.

(g) In sloped or stepped footings, the supporting area for bearing may be taken as the top horizontal surface of the footing, or assumed as the area of the lower base of the largest frustum of a pyramid or cone contained wholly within the footing and having for its upper base the area actually loaded, and having side slopes of one vertical to two horizontal.

2307—Pedestals and footings (plain concrete)

(a) The allowable compression stress on the gross area of a concentrically loaded pedestal under service load shall not exceed $0.25f_c'$. Where this stress is exceeded, reinforcement shall be provided and the member designed as a reinforced concrete column.

(b) The depth and width of a pedestal or footing of plain concrete shall be such that the tension in the concrete in flexure shall not exceed $1.6\sqrt{f_c'}$ for design by Part IV-A or $3.2\sqrt{f_c'}$ for design by Part IV-B. The average shear stress shall satisfy the requirements of Chapter 12 or 17.

2308—Footings supporting round columns

(a) In computing the stresses in footings which support a round or octagonal concrete column or pedestal, the "face" of the column or pedestal may be taken as the side of a square having an area equal to the area enclosed within the perimeter of the column or pedestal.

2309—Minimum edge thickness

(a) In reinforced concrete footings, the thickness above the reinforcement at the edge shall be not less than 6 in. for footings on soil, nor less than 12 in. for footings on piles.

(b) In plain concrete footings, the thickness at the edge shall be not less than 8 in. for footings on soil, nor less than 14 in. above the tops of the piles for footings on piles.

2310—Combined footings and mats

(a) The following recommendations are made for combined footings and mats — those supporting more than one column or wall:

1. Soil pressures shall be considered as acting uniformly or varying linearly, except that other assumptions may be made consistent with the properties of the soil and the structure and with established principles of soil mechanics.
2. Shear as a measure of diagonal tension shall be computed in conformance with Section 1207 or 1707.

A.6 Building Code Requirements for Foundations

The material on the following pages consists of selected reprints from the *City of Los Angeles Building Code* and the *Uniform Building Code*. While most of this material is similar to that found in any building code, local situations and practices produce unique requirements and recommendations in most city, county and state codes.

Excerpts from the *City of Los Angeles Building Code*, 1976 edition, are reprinted with permission of the publisher, Building News, Inc., 3055 Overland Avenue, Los Angeles, CA 90034.

Excerpts from the *Uniform Building Code*, 1979 edition, copyright 1979, are reprinted with permission of the publisher, International Conference of Building Officials, 5360 South Workman Mill Road, Whittier, CA 90601.

A.7 Excerpts from *City of Los Angeles Building Code*

SEC. 91.2309 — RETAINING WALLS

(a) **Design.** Retaining walls shall be designed to resist the lateral pressure of the retained material determined in accordance with accepted engineering principles.

The soil characteristics and design criteria necessary for such a determination shall be obtained from a special foundation investigation performed by an agency acceptable to the Department. The Department shall approve such characteristics and criteria only after receiving a written opinion from the investigation agency together with substantiating evidence.

EXCEPTION: Freestanding walls which are not over 15' in height or basement walls which have spans of 15' or less between supports may be designed in accordance with Subsection (b) of this Section.

TABLE NO. 23-E

Surface Slope of Retained Material* Horiz. to Vert.	Equivalent Fluid Weight lb/ft³
LEVEL	30
5 to 1	32
4 to 1	35
3 to 1	38
2 to 1	43
1½ to 1	55
1 to 1	80

* Where the surface slope of the retained earth varies, the design slope shall be obtained by connecting a line from the top of the wall to the highest point on the slope, whose limits are within the horizontal distance from the stem equal to the stem height of the wall.

(b) **Arbitrary Design Method.** Walls which retain drained earth and come within the limits of the exception to Subsection (a) of this section may be designed for an assumed earth pressure equivalent to that exerted by a fluid weighing not less than shown in Table 23-E. A vertical component equal to one-third of the horizontal force so obtained may be assumed at the plane of application of the force.

The depth of the retained earth shall be the vertical distance below the ground surface measured at the wall face for stem design or measured at the heel of the footing for overturning and sliding.

(c) **Surcharge.** Any superimposed loading, except retained earth, shall be considered as surcharge and provided for in the design. Uniformly distributed loads may be considered as equivalent added depth of retained earth. Surcharge loading due to continuous or isolated footings shall be determined by the following formulas or by an equivalent method approved by the Superintendent of Building.

Resultant Lateral Force

$$R = \frac{0.3\,Ph^2}{x^2 + h^2}$$

Location of Lateral Resultant

$$d = x\left[\left(\frac{x^2}{h^2} + 1\right)\left(\tan^{-1}\frac{h}{x}\right) - \left(\frac{x}{h}\right)\right]$$

Where:

R = Resultant lateral force measured in pounds per foot of wall width.

P = Resultant surcharge load of continuous or isolated footings measured in pounds per foot of length parallel to the wall.

x = Distance of resultant load from back face of wall measured in feet.

h = Depth below point of application of surcharge loading to top of wall footing measured in feet.
d = Depth of lateral resultant below point of application of surcharge loading measured in feet.

$\left(\tan^{-1}\frac{h}{x}\right)$ = The angle in radians whose tangent is equal to $\left(\frac{h}{x}\right)$

Loads applied within a horizontal distance equal to the wall stem height, measured from the back face of the wall, shall be considered as surcharge.

For isolated footings having a width parallel to the wall less than three feet, "R" may be reduced to 1/6 the calculated value.

The resultant lateral force "R" shall be assumed to be uniform for the length of footing parallel to the wall, and to diminish uniformly to zero at the distance "x" beyond the ends of the footing.

Vertical pressure due to surcharge applied to the top of the wall footing may be considered to spread uniformly within the limits of the stem and planes making an angle of 45° with the vertical.

(d) **Bearing Pressure and Overturning.** The maximum vertical bearing pressure under any retaining wall shall not exceed that allowed in Division 28 of this Article except as provided for by a special foundation investigation. The resultant of vertical loads and lateral pressures shall pass through the middle one-third of the base.

(e) **Friction and Lateral Soil Pressures.** Retaining walls shall be restrained against sliding by friction of the base against the earth, by lateral resistance of the soil, or by a combination of the two. Allowable friction and lateral soil values shall not exceed those allowed in Division 28 of this Article except as provided by a special foundation investigation.

When used, keys shall be assumed to lower the plane of frictional resistance and the depth of lateral bearing to the level of the bottom of the key. Lateral bearing pressures shall be assumed to act on a vertical plane located at the toe of the footing.

(f) **Construction.** No retaining wall shall be constructed of wood.

(g) **Special Conditions.** Whenever, in the opinion of the Superintendent of Building, the adequacy of the foundation material to support a wall is questionable, an unusual surcharge condition exists, or whenever the retained earth is so stratified or of such a character as to invalidate normal design assumptions, he may require a special foundation investigation before approving any permit for such a wall.

SEC. 91.2310 — TRUSSES, ARCHES AND GIRDERS

Trusses, arches, and girders having a span greater than 40 feet, and supported by masonry or concrete columns or walls shall have a length of bearing, measured parallel to the span, of not less than eight inches.

> ***EXCEPTION:*** *Lesser bearing lengths will be permitted when the trusses, arches, or girders are fabricated using actual measured distances between bearings, or when other approved procedures are used which will assure bearing centered upon the line used in the design calculations.*

SEC. 91.2311 — POLES

(a) **Design.** Flag poles, sign poles, columns or other poles cantilevering from and receiving lateral stability from the ground shall have their lateral support designed in accordance with the following formulas or other methods approved by the Superintendent of Building. Bearing stresses so obtained shall not exceed the values permitted by Section 91.2803 (d).

CASE I — POLES WITH LATERAL RESTRAINT AT THE GROUND SURFACE

$$f = \frac{3.8M}{bd^2}$$

Where:

f = lateral soil pressure in lbs/sq. ft.

M = moment at natural ground surface resulting from applied loads in ft. pounds.

b = diameter of round pole or 1.27 times width of rectangular pole, measured in feet.

R = Reaction capable of taking resultant loads.

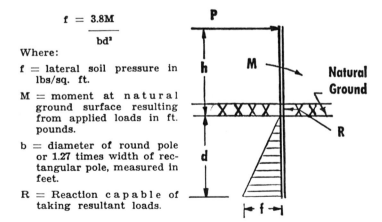

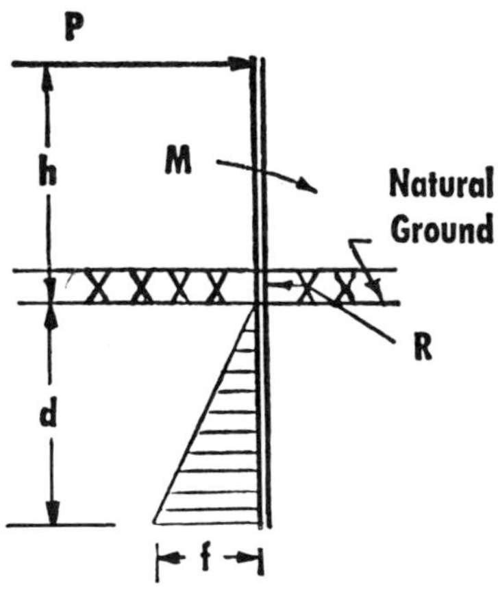

CASE II — POLES WITHOUT LATERAL RESTRAINT AT THE GROUND SURFACE

$$f_1 = \frac{2.85\ P}{bd} + \frac{f_2}{4}$$

$$f_2 = \frac{7.62\ P\ (2h + d)}{bd^3}$$

Where:

f_1 and f_2 = lateral soil pressure in lbs/sq. ft.

b = diameter of round pole or 1.27 times width of rectangular pole, measured in feet.

d = depth of embedment below natural ground in feet (minimum four feet).

h = height of applied lateral load above natural ground measured in feet.

P = lateral force in pounds.

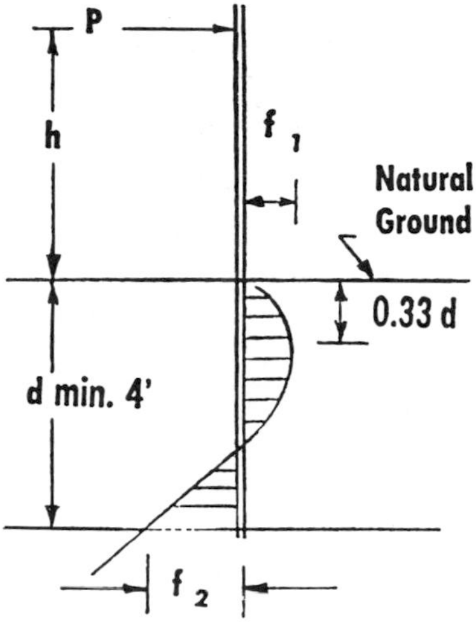

TABLE NO. 28-A — ALLOWABLE FOUNDATION PRESSURE
(Kips per Square Foot — 1 Kip = 1,000 pounds)

CLASS OF MATERIAL Rock—Depth of Embedment shall be to a Fresh Unweathered Surface Except as Noted	Value at Min. Depth	Increase for Depth	Maximum Value
***Massive crystalline bedrock;** basalt, granite and diorite in sound condition	20		20
***Foliated rocks;** schist and slate, in sound condition	8		8
***Sedimentary rocks;** hard shales, dense siltstones and sandstones, thoroughly cemented conglomerates	6		6
Soft, or broken bedrocks; soft shales, shattered slates, diatomaceous shales; other badly jointed (fractured) or weathered rock. 12" minimum embedment	2		2

***NOTE:** The above values apply only where the strata are level or nearly so, and/or where the area has ample lateral support. Tilted strata, and the relationship to nearby slopes should receive special consideration. These values may be increased one-third to a maximum of two times the assigned value, for each foot of penetration below fresh, unweathered surface.

TABLE NO. 28-A (Continued)

Soils—Minimum Depth of Embedment shall be one foot below the adjacent undisturbed ground surface*	Loose	Compact	Soft	Stiff	Increase for Depth	Maximum Value
Gravel, well graded. Well graded gravels or gravel-sand mixtures, little or no fines	1.33	2.0			20	8
Gravel, poorly graded. Poorly graded gravels or gravel-sand mixtures, little or no fines	1.33	2.0			20	8
Gravel, silty. Silty gravels or poorly graded gravel sand silt mixtures	1.0	2.0			20	8
Gravel, clayey. Clayey gravels or gravel-sand clay mixtures	1.0	2.0			20	8
Sand, well graded. Well graded sands or gravelly sands, little or no fines	1.0	2.0			20	6
Sand, poorly graded. Poorly graded sands or gravelly sands, little or no fines	1.0	2.0			20	6
Sand, silty. Silty sand, or poorly graded sand-silt mixtures	0.5	1.5			20	4
Sand, clayey. Clayey sands or sand-clay mixtures	1.0	2.0			20	4
Silt. Inorganic silts and very fine sands, rock flour, silty or clayey fine sands with slight plasticity	0.5	1.0			20	3
Silt, organic. Organic silts and organic silt-clays of low plasticity	0.5	1.0	0.5	1.0	10	2
Silt, elastic. Very compressible silts, micaceous or diatomaceous fine sandy or silty soils	0.5	1.0			10	1.5
Clay, lean. Inorganic clays of low to medium plasticity, silty clays, lean clays	1.0	2.0	1.0	2.0	20	3
Clay, fat. Very compressible clays, inorganic clays of high plasticity			0.5	1.0	10	1.5
Clay, organic. Organic clays of medium to high plasticity, very compressible			0.5			0.5
Peat. Peat and other highly organic swamp soils			0			0

NOTES:

1. Values for gravels and sand given are for footings one foot in width and may be increased in direct proportion to footing width to maximum of three times the maximum value, or to the designated maximum value, whichever is the least.
2. Where the bearing values in the above table are used, it should be noted that increased width or unit load will cause increase in settlement.
3. Special attention should be given to the effect of increase in moisture in establishing soil classifications.

*4. Minimum depth for highly expansive soils to be one and one-half feet.

5. Increases for depth are given in percentage of minimum value for each additional foot below the minimum required depth.

TABLE NO. 28-B — ALLOWABLE FRICTIONAL & BEARING VALUES FOR ROCK[1]

Type	Friction Coefficient	Allowable Lateral Bearing lbs. per sq. ft.	Allowable Lateral Bearing per ft. Max. Value
Massive Crystalline Bedrock	1.0	4,000	20,000
Foliated Rocks	.8	1,600	8,000
Sedimentary Rocks	.6	1,200	6,000
Soft or Broken Bedrocks	.4	400	2,000

TABLE NO. 28-B (Continued)

ALLOWABLE FRICTIONAL & LATERAL BEARING VALUES FOR SOILS

Frictional Resistance — Gravels and Sands[1]

Soil Type	Friction Coefficient
Gravel, Well Graded	0.6
Gravel, Poorly Graded	0.6
Gravel, Silty	0.5
Gravel, Clayey	0.5
Sand, Well Graded	0.4
Sand, Poorly Graded	0.4
Sand, Silty	0.4
Sand, Clayey	0.4

1. Coefficient to be multiplied by the Dead Load.

ALLOWABLE FRICTIONAL RESISTANCE (lbs. per sq. ft.) — Clay and Silt[2]

Soil Type	Loose or Soft	Compact or Stiff
Silt, Inorganic	250	500
Silt, Organic	250	500
Silt, Elastic	200	400
Clay, Lean	500	1000
Clay, Fat	200	400
Clay, Organic	150	300
Peat	0	0

2. Frictional values to be multiplied by the width of footing subjected to positive soil pressure. In no case shall the fricitional resistance exceed ½ the dead load on the area under consideration.

ALLOWABLE LATERAL BEARING PER FT. OF DEPTH BELOW NATURAL GROUND SURFACE (lbs. per sq. ft.) (Natural Soils or approved compacted fill)

Soil Type	Loose or Soft	Compact or Stiff	Max. Values
Gravel, Well Graded	200	400	8000
Gravel, Poorly Graded	200	400	8000
Gravel, Silty	167	333	8000
Gravel, Clayey	167	333	8000
Sand, Well Graded	183	367	6000
Sand, Poorly Graded	77	200	6000
Sand, Silty	100	233	4000
Sand, Clayey	133	300	4000
Silt, Inorganic	67	133	3000
Silt, Organic	33	67	2000
Silt, Elastic	33	67	1500
Clay, Lean	267	667	3000
Clay, Fat	33	167	1500
Clay, Organic	33	......	500
Peat	0	0	0

GENERAL CONDITIONS OF USE

1. Frictional and lateral resistance of soils may be combined, provided the lateral bearing resistance does not exceed ⅔ of allowable lateral bearing.

2. A ⅓ increase in frictional and lateral bearing values will be permitted to resist loads caused by wind pressure or earthquake forces.

3. Isolated poles such as flag poles or signs may be designed using lateral bearing values equal to two times the tabulated values.

4. Lateral bearing values are permitted only when concrete is deposited against natural ground or compacted fill, approved by the Superintendent of Building.

A.8 Excerpts from *Uniform Building Code*

Chapter 29

EXCAVATIONS, FOUNDATIONS AND RETAINING WALLS

Scope

Sec. 2901. This chapter sets forth requirements for excavation and fills for any building or structure and for foundations and retaining structures.

Reference is made to Appendix Chapter 70 for requirements governing excavation, grading and earthwork construction, including fills and embankments.

Quality and Design

Sec. 2902. The quality and design of materials used structurally in excavations, footings and foundations shall conform to the requirements specified in Chapters 23, 24, 25, 26 and 27 of this code.

Excavations and Fills

Sec. 2903. (a) **General.** Excavation or fills for buildings or structures shall be so constructed or protected that they do not endanger life or property.

Cut slopes for permanent excavations shall not be steeper than 2 horizontal to 1 vertical and slopes for permanent fills shall not be steeper than 2 horizontal to 1 vertical unless substantiating data justifying steeper slopes are submitted. Deviation from the foregoing limitations for slopes shall be permitted only upon the presentation of a soil investigation report acceptable to the building official.

No fill or other surcharge loads shall be placed adjacent to any building or structure unless such building or structure is capable of withstanding the additional loads caused by the fill or surcharge.

Existing footings or foundations which may be affected by any excavation shall be underpinned adequately or otherwise protected against settlement and shall be protected against lateral movement.

Fills to be used to support the foundations of any building or structure shall be placed in accordance with accepted engineering practice. A soil investigation report and a report of satisfactory placement of fill, both acceptable to the building official, shall be submitted.

(b) **Protection of Adjoining Property.** The requirements for protection of adjacent property and depth to which protection is required shall be as defined by prevailing law. Where not defined by law, the following shall apply: Any person making or causing an excavation to be made to a depth of 12 feet or less below the grade shall protect the excavation so that the soil of adjoining property will not cave in or settle, but shall not be liable for the expense of underpinning or extending the foundation of buildings on adjoining properties where his excavation is not in excess of 12 feet in depth. Before commencing the excavation, the person making or causing the excavation to be made shall notify in writing the owners of adjoining buildings not less than 10 days before such excavation is to be made that the excavation is to be made and that the adjoining buildings should be protected.

The owners of the adjoining properties shall be given access to the excavation for the purpose of protecting such adjoining buildings.

Any person making or causing an excavation to be made exceeding 12 feet in depth below the grade shall protect the excavation so that the adjoining soil will not cave in or settle and shall extend the foundation of any adjoining buildings below the depth of 12 feet below grade at his own expense. The owner of the adjoining buildings shall extend the foundation of these buildings to a depth of 12 feet below grade at his own expense, as provided in the preceding paragraph.

Soil Classification—Expansive Soil

Sec. 2904. (a) **Soil Classification: General.** For the purposes of this chapter, the definition and classification of soil materials for use in Table No. 29-B shall be according to U.B.C. Standard No. 29-1.

(b) **Expansive Soil.** When the expansive characteristics of a soil are to be determined, the procedures shall be in accordance with U.B.C. Standard No. 29-2 and the soil shall be classified according to Table No. 29-C. Foundations for structures resting on soils with an expansion index greater than 20, as determined by U.B.C. Standard No. 29-2, shall require special design consideration. In the event the soil expansion index varies with depth, the weighted index shall be determined according to Table No. 29-D.

Foundation Investigation

Sec. 2905. (a) **General.** The classification of the soil at each building site shall be determined when required by the building official. The building official may require that this determination be made by an engineer or architect licensed by the state to practice as such.

(b) **Investigation.** The classification shall be based on observation and any necessary tests of the materials disclosed by borings or excavations made in appropriate locations. Additional studies may be necessary to evaluate soil strength, the effect of moisture variation on soil bearing capacity, compressibility and expansiveness.

(c) **Reports.** The soil classification and design bearing capacity shall be shown on the plans, unless the foundation conforms to Table No. 29-A. The building official may require submission of a written report of the investigation which shall include, but need not be limited to, the following information:

1. A plot showing the location of all test borings and/or excavations.
2. Descriptions and classifications of the materials encountered.

3. Elevation of the water table, if encountered.
4. Recommendations for foundation type and design criteria including bearing capacity, provisions to minimize the effects of expansive soils and the effects of adjacent loads.
5. Expected total and differential settlement.

(d) **Expansive Soils.** When expansive soils are present, the building official may require that special provisions be made in the foundation design and construction to safeguard against damage due to this expansiveness. He may require a special investigation and report to provide this design and construction criteria.

(e) **Adjacent Loads.** Where footings are placed at varying elevations the effect of adjacent loads shall be included in the foundation design.

(f) **Drainage.** Provisions shall be made for the control and drainage of surface water around buildings.

Allowable Foundation and Lateral Pressures

Sec. 2906. The allowable foundation and lateral pressures shall not exceed the values set forth in Table No. 29-B unless data to substantiate the use of higher values are submitted. Table No. 29-B may be used for design of foundations on rock or nonexpansive soil for Types II One-hour, II-N and V buildings which do not exceed three stories in height or for structures which have continuous footings having a load of less than 2000 pounds per lineal foot and isolated footings with loads of less than 50,000 pounds.

Footings

Sec. 2907. (a) **General.** Footings and foundations, unless otherwise specifically provided, shall be constructed of masonry, concrete or treated wood in conformance with U.B.C. Standard No. 29-3 and in all cases shall extend below the frost line. Footings of concrete and masonry shall be of solid material. Foundations supporting wood shall extend at least 6 inches above the adjacent finish grade. Footings shall have a minimum depth below finished grade as indicated in Table No. 29-A unless another depth is recommended by a foundation investigation.

(b) **Bearing Walls.** Bearing walls shall be supported on masonry or concrete foundations or piles or other approved foundation system which shall be of sufficient size to support all loads. Where a design is not provided, the minimum foundation requirements for stud bearing walls shall be as set forth in Table No. 29-A.

> **EXCEPTIONS:** 1. A one-story wood or metal frame building not used for human occupancy and not over 400 square feet in floor area may be constructed with walls supported on a wood foundation plate when approved by the building official.
>
> 2. The support of buildings by posts embedded in earth shall be designed as specified in Section 2907 (f). Wood posts or poles embedded in earth shall be pressure treated with an approved preservative. Steel posts or poles shall be protected as specified in Section 2908 (h).

(c) **Stepped Foundations.** Foundations for all buildings where the surface of the ground slopes more than 1 foot in 10 feet shall be level or shall be stepped so that both top and bottom of such foundation are level.

(d) **Footing Design.** Except for special provisions of Section 2909 covering the design of piles, all portions of footings shall be designed in accordance with the structural provisions of this code and shall be designed to minimize differential settlement.

(e) **Foundation Plates or Sills.** Foundation plates or sills shall be bolted to the foundation or foundation wall with not less than ½-inch-diameter steel bolts embedded at least 7 inches into the concrete or reinforced masonry or 15 inches into unreinforced grouted masonry and spaced not more than 6 feet apart. There shall be a minimum of two bolts per piece with one bolt located within 12 inches of each end of each piece. Foundation plates and sills shall be the kind of wood specified in Section 2517 (c).

(f) **Designs Employing Lateral Bearing.** Construction employing posts or poles as columns embedded in earth or embedded in concrete footings in the earth may be used to resist both axial and lateral loads. The depth to resist lateral loads shall be determined by means of the design criteria established herein or other methods approved by the building official.

1. **Design criteria: Nonconstrained.** The following formula may be used in determining the depth of embedment required to resist lateral loads where no constraint is provided at the ground surface, such as rigid floor or rigid ground surface pavement:

$$d = \frac{A}{2}\left(1 + \sqrt{1 + \frac{4.36h}{A}}\right)$$

WHERE:

$A = \dfrac{2.34P}{S_1 b}$

P = Applied lateral force in pounds.

S_1 = Allowable lateral soil-bearing pressure as set forth in Table No. 29-B based on a depth of one-third the depth of embedment.

S_3 = Allowable lateral soil-bearing pressure as set forth in Table No. 29-B based on a depth equal to the depth of embedment.

b = Diameter of round post or footing or diagonal dimension of square post or footing (feet).

h = Distance in feet from ground surface to point of application of "P".

d = Depth of embedment in earth in feet but not over 12 feet for purpose of computing lateral pressure.

Constrained. The following formula may be used to determine the depth of embedment required to resist lateral loads where constraint is provided at the ground surface, such as a rigid floor or pavement:

$$d^2 = 4.25 \frac{Ph}{S_3 b}$$

Vertical Load. The resistance to vertical loads is determined by the allowable soil-bearing pressure set forth in Table No. 29-B.

2. **Construction requirements: Backfill.** The backfill in the annular space around columns not embedded in poured footings shall be by one of the following methods:

A. Backfill shall be of concrete with an ultimate strength of 2000 pounds per square inch at 28 days. The hole shall be not less than 4 inches larger than the diameter of the column at its bottom or 4 inches larger than the diagonal dimension of a square or rectangular column.

B. Backfill shall be of clean sand. The sand shall be thoroughly compacted by tamping in layers not more than 8 inches in depth.

3. **Limitations.** The design procedure outlined in this subsection shall be subject to the following limitations:

The frictional resistance for retaining walls and slabs on silts and clays shall be limited to one-half of the normal force imposed on the soil by the weight of the footing or slab.

Posts embedded in earth shall not be used to provide lateral support for structural or nonstructural materials such as plaster, masonry or concrete unless bracing is provided that develops the limited deflection required.

(g) **Grillage Footings.** When grillage footings of structural steel shapes are used on soils, they shall be completely embedded in concrete with at least 6 inches on the bottom and at least 4 inches at all other points.

(h) **Bleacher Footings.** Footings for open air seating facilities shall comply with Chapter 29.

> **EXCEPTION:** Temporary open air portable bleachers as defined in Section 3321 may be supported upon wood sills or steel plates placed directly upon the ground surface, provided soil pressure does not exceed 1200 pounds per square foot.

TABLE NO. 29-A—FOUNDATIONS FOR STUD BEARING WALLS—MINIMUM REQUIREMENTS

NUMBER OF STORIES	THICKNESS OF FOUNDATION WALL (Inches)		WIDTH OF FOOTING (Inches)	THICKNESS OF FOOTING (Inches)	DEPTH OF FOUNDATION BELOW NATURAL SURFACE OF GROUND AND FINISH GRADE (Inches)
	CONCRETE	UNIT MASONRY			
1	6	6	12	6	12
2	8	8	15	7	18
3	10	10	18	8	24

NOTES:

Where unusual conditions or frost conditions are found, footings and foundations shall be as required in Section 2907 (a).

The ground under the floor may be excavated to the elevation of the top of the footing.

TABLE NO. 29-C—CLASSIFICATION OF EXPANSIVE SOIL

EXPANSION INDEX	POTENTIAL EXPANSION
0-20	Very Low
21-50	Low
51-90	Medium
91-130	High
Above 130	Very High

TABLE NO. 29-D—WEIGHTED EXPANSION INDEX

DEPTH INTERVAL[2]	WEIGHT FACTOR
0-1	0.4
1-2	0.3
2-3	0.2
3-4	0.1
Below 4	0

[1]The weighted expansion index for non-uniform soils is determined by multiplying the expansion index for each depth interval by the weight factor for that interval and summing the products.

[2]Depth in feet below the ground surface.

TABLE NO. 29-B—ALLOWABLE FOUNDATION AND LATERAL PRESSURE

CLASS OF MATERIALS[2]	ALLOWABLE FOUNDATION PRESSURE LBS. SQ. FT.[3]	LATERAL BEARING LBS./SQ. FT./FT. OF DEPTH BELOW NATURAL GRADE[4]	LATERAL SLIDING[1]	
			COEFFICIENT[5]	RESISTANCE LBS./SQ. FT.[6]
1. Massive Crystalline Bedrock	4000	1200	.79	
2. Sedimentary and Foliated Rock	2000	400	.35	
3. Sandy Gravel and/or Gravel (GW and GP)	2000	200	.35	
4. Sand, Silty Sand, Clayey Sand, Silty Gravel and Clayey Gravel (SW, SP, SM, SC, GM and GC)	1500	150	.25	
5. Clay, Sandy Clay, Silty Clay and Clayey Silt (CL, ML, MH and CH)	1000[7]	100		130

[1]Lateral bearing and lateral sliding resistance may be combined.

[2]For soil classifications OL, OH and Pt (i.e., organic clays and peat), a foundation investigation shall be required.

[3]All values of allowable foundation pressure are for footings having a minimum width of 12 inches and a minimum depth of 12 inches into natural grade. Except as in Footnote 7 below, increase of 20 percent allowed for each additional foot of width and/or depth to a maximum value of three times the designated value.

[4]May be increased the amount of the designated value for each additional foot of depth to a maximum of 15 times the designated value. Isolated poles for uses such as flagpoles or signs and poles used to support buildings which are not adversely affected by a ½-inch motion at ground surface due to short-term lateral loads may be designed using lateral bearing values equal to two times the tabulated values.

[5]Coefficient to be multiplied by the dead load.

[6]Lateral sliding resistance value to be multiplied by the contact area. In no case shall the lateral sliding resistance exceed one-half the dead load.

[7]No increase for width is allowed.

Answers to Selected Exercise Problems

Chapter 2

1A. Express volumes in terms of proportionate parts of a total volume of one. Thus

$$\text{total volume} = (\text{volume of solids}) + (\text{volume of void})$$

$$V_{tot} = V_s + V_v = 1$$

Unit dry weight is actually the weight of the solids.

$$\gamma = (V_s)(G)(62.4) = 90$$

Using the assumed G of 2.65,

$$V_s = \frac{90}{(2.65)(62.4)} = 0.544$$

and

$$V_v = 1 - V_s = 1 - 0.544 = 0.456$$

Then

$$e = \frac{V_v}{V_s} = \frac{0.456}{0.544} = \underline{\underline{0.838}}$$

If sample is saturated,

$$\text{weight of water} = (V_v)(62.4) = (0.456)(62.4) = 28 \text{ lb}$$

Total saturated weight

$$\gamma = 90 + 28 = \underline{\underline{118 \text{ lb/ft}^3}}$$

If the total sample weight is 110 lb/ft³, the weight of the water in the sample is 110 minus the weight of the solids. Thus

$$W_s = 110 - 90 = 20 \text{ lb}$$

Water content

$$w = \frac{W_w}{W_s}(100) = \frac{20}{90}(100) = \underline{\underline{22\%}}$$

2A. For the saturated sample, with the void completely filled with water,

$$\gamma = (\text{weight of solids}) + (\text{weight of water})$$

$$90 = (V_s)(G)(62.4) + (V_v)(62.4) = (62.4)(2.7\,V_s + V_v)$$

and, since $V_s + V_v = 1$,

$$90 = 62.4[2.7\,V_s + (1 - V_s)] = 62.4(1.7\,V_s + 1)$$

from which

$$V_s = 0.260,\ V_v = 1 - V_s = 0.740$$

For the dry weight,

$$W_s = (0.260)(2.7)(62.4) = \underline{\underline{43.8 \text{ lb/ft}^3}}$$

Weight of the water

$$W_w = (0.740)(62.4) = 46.2 \text{ lb/ft}^3$$

For a check,

$$W_s + W_w = 43.8 + 46.2 = 90 \text{ lb/ft}^3 \text{ (as given)}$$

Void ratio

$$e = \frac{V_v}{V_s} = \frac{0.740}{0.260} = \underline{\underline{2.85}}$$

Water content

$$w = \frac{W_w}{W_s} = \frac{46.2}{43.8}(100) = \underline{\underline{105\%}}$$

3A. Sketch the line for the size gradation on the graph in text Figure 2.2 to classify the general type of soil. Sample A is 70% gravel and has less than 10% fines, and would thus be called a sandy gravel. To classify the type of gradation determine the values for C_u and C_z.

$$C_u = \frac{D_{60}}{D_{10}} = \frac{50}{1} = 50$$

$$C_z = \frac{(D_{30})^2}{(D_{10})(D_{60})} = \frac{(4)^2}{(1)(50)} = 0.32$$

Because C_z is less than one, the soil is considered poorly graded.

4A. With 60% of the particles retained on the No. 4 sieve this sample is a gravel (GW, GP, GM, or GC). With less than 5% passing the No. 200 sieve (constituting fines) it is a clean gravel (GW or GP).

$$C_u = \frac{D_{60}}{D_{10}} = \frac{8}{0.4} = 20$$

$$C_z = \frac{(D_{30})^2}{(D_{10})(D_{60})} = \frac{(3)^2}{(0.4)(8)} = 2.81$$

which qualifies it as well graded, or GW.
With the N of 20, Table 2.5 qualifies it as medium dense and gives the following approximate values:

Allowable bearing: 1500 lb/ft^2 + 200 lb/ft^2 for each foot of surcharge over 1.5.

Passive resistance: 300 lb/ft of depth below grade.

Friction: 0.6 times the dead load.

Chapter 3

1. Table entries should be reasonably correct. In many cases footing thicknesses are not limited by concrete strength, but have been increased to reduce the amount of reinforcing.

2. Strength design methods will in some instances result in minor savings of concrete and/or reinforcing. This depends somewhat on the proportions of the live and dead loads, since this variable is incorporated in the strength design methods.

3. Since the footing thickness is usually increased for the oblong footing, whereas the required plan area remains approximately the same, there is ordinarily some increased concrete volume. The total weight of reinforcing, however, should remain approximately the same.

4A. For the 100 k load, Table 3.5 gives a footing 6 ft square and 16 in. thick, which weighs approximately 7 k. If we add a footing weight for the combined footing of approximately 15 k to the sum of the column loads, we may determine the total footing area required.

$$A = \frac{215}{3} = 71.67 \text{ ft}^2$$

Various combinations of width and length may be used to obtain this area, such as 10 by 7.17, 11 by 6.52, 12 by 5.97, 13 by 5.51. Any of these will work, although I favor the 12 ft or longer footing, since that will eliminate the need for reinforcing in the top of the footing.

5A. The total footing area required is first estimated by adding a value for the weight of the footing to the column loads and dividing by the allowable soil pressure.

$$A = \frac{325}{3} = 108.3 \text{ ft}^2$$

The next decision is whether to use a rectangular or T-shaped plan. If we choose a rectangular plan, we find the desired length by determining the centroid of the loads, as shown in the example in the text. Determining a T-shape is somewhat more complex, and will most likely involve trying several combinations of areas in order to find one that has the proper total area, location of centroid, and reasonable size of parts for the structural actions.

6A. Assuming a 6 in. thick wall, an 8 in. thick footing, and a soil weight of 100 lb/ft^3, a 20 in. wide footing will just keep the resultant inside the kern. Sliding and overturn will not be critical in this case.

7A. Determination of live load to dead load ratios will show that Column C is the critical footing. For the total load of 74 k we find from Table 3.5 that a footing will be approximately 6.5 ft square and 14 in. thick for this column. If this footing is used the dead load pressure will be

$$p = \frac{26{,}000}{(6.5)^2} + \text{the weight of the footing}$$

$$= 615 + 175 = 790 \text{ lb/ft}^2$$

which is used to determine the required areas for the other footings. This will produce the approximate results as follows:

Wall A: 20 in. wide by 6 in. thick, no transverse reinforcing.
Wall B: 32 in. wide by 10 in. thick, no transverse reinforcing.
Column D: 9 ft square by 18 in. thick, 10 No. 7 bars each way.

Index